Boulevard of Dreams

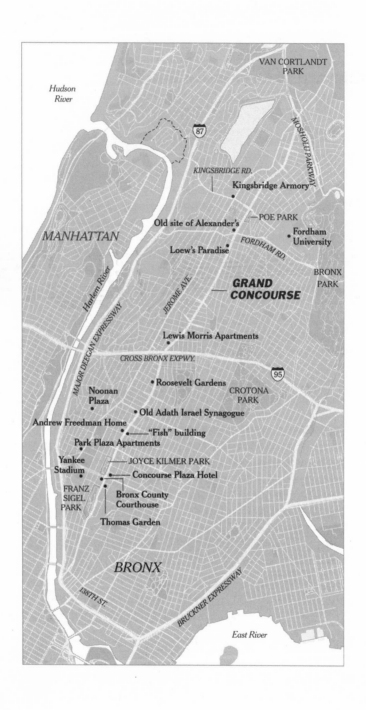

Boulevard of Dreams

*Heady Times, Heartbreak, and Hope
along the Grand Concourse in the Bronx*

Constance Rosenblum

NEW YORK UNIVERSITY PRESS
New York and London

NEW YORK UNIVERSITY PRESS
New York and London
www.nyupress.org

Frontispiece: Courtesy Natasha Perkel.

Library of Congress Cataloging-in-Publication Data

Rosenblum, Constance.
 Boulevard of dreams : heady times, heartbreak, and hope along
the Grand Concourse in the Bronx / Constance Rosenblum.
p. cm.
Includes bibliographical references and index.
ISBN-13: 978-0-8147-7608-7 (cl : alk. paper)
ISBN-10: 0-8147-7608-6 (cl : alk. paper)
1. Grand Concourse (New York, N.Y.)—History. 2. Bronx (New York, N.Y.)—History.
3. Bronx (New York, N.Y.)—Social conditions. 4. Bronx (New York, N.Y.)—Social life
and customs. 5. Jews—New York (State)—New York—History. I. Title.
F128.68.B8R67   2009
974.7'275—dc22   2009011262

Manufactured in the United States of America

10 9 8 7 6 5 4 3 2 1

To Andy and Sarah

Contents

Acknowledgments

THIS BOOK EXISTS because countless people were kind enough to share their insights and expertise with me over the years.

I'd especially like to thank Sam Goodman, Mark Caldwell, Thomas Mellins, Lloyd Ultan, Evelyn Gonzalez, Mark Naison, Jerome Charyn, Marshall Berman, Leonard Kriegel, Arthur Gelb, Avery Corman, James Crocker, Gelvin Stevenson, Deborah Dash Moore, and Robert Caro, all of whom were exceptionally generous with their time and did a great deal to help clarify my thinking about the evolving world of the Grand Concourse.

I am indebted beyond words to Stephen Sinon, head of information services and archives at the LuEsther T. Mertz Library of the New York Botanical Garden, who served as my research assistant. Stephen not only unearthed every document I asked for, he also directed me to countless others that made this book richer and more informative. And given his yeoman work collecting images and acquiring the sometimes elusive rights to them, he is the one most responsible for the illustrations.

I am also deeply indebted to those who were kind enough to critique all or parts of my manuscript and offered invaluable suggestions; in addition to Mark Caldwell, Sam Goodman, Thomas Mellins, and Lloyd Ultan, they include Jim Rasenberger, Manette Berlinger, and my *New York Times* colleagues Mitch Keller and John Oudens.

I'm grateful to Mimi Vang Olsen for sharing the story of her father, Kourken Hovsepian, and his priceless photographs of generations of West Bronx families, a collection that I hope someday finds a public home of its own. I'm also grateful to Robert Billingsley for sharing documents and memories about his father, the developer Logan Billingsley; to John Ginsbern for memories of his grandfather, the architect Horace Ginsbern; to Sonia and Paula Kessler for recollections about the poet Milton Kessler; to Suzanne Callahan and Maury Brassert for family memories about the theater designer John Eberson; and to James Crocker for sharing a scrapbook about Andrew Freedman that provided a vivid portrait of a man, an institution, and an era.

At the Bronx African-American History Project, in addition to Mark Naison, I'd like to thank Brian Purnell and the individuals whose descriptions of life in the borough enriched my understanding of Bronx life, among them Leroi Archible, Beatrice Bergland, Jesse Davidson, Joan Tyson Fortune, Robert Gumbs, Allen Jones, and Cyril DeGrasse Tyson.

At the Bronx borough president's office, in addition to Sam Goodman, I'd like to thank Daniel Donovan and Wilhelm Ronda. I'm also grateful to three former borough presidents—Robert Abrams, Fernando Ferrer, and Herman Badillo—all of whom shared their insights about the strengths and troubles of the West Bronx, as did current and former staff members of the New York City Planning Department, among them Robert Esnard, Lloyd Kaplan, Larry Parnes, and Rachaele Raynoff.

Staff at libraries and other institutions around the city went far beyond the call of duty in helping me with my research. I'd especially like to thank Laura Tosi, Peter Derrick, and Gary Hermalyn at the Bronx County Historical Society; Janet Munch at the Leonard Lief Library of Lehman College of the City University of New York, and her colleague William Bosworth, director of Lehman's Bronx Data Center; Melanie Bower at the Museum of the City of New York; Janet Parks at Avery Architectural and Fine Arts Library at Columbia University, home of the Horace Ginsbern archives; Sergio Bessa, Holly Block, and their colleagues at the Bronx Museum of the Arts; William Casari at Hostos Community College; Miranda Schwartz at the New-York Historical Society; and staff members at the YIVO Institute for Jewish Research, the Center for Puerto Rican Studies at Hunter College, the American Museum of the Moving Image, and the Paley Center for Media, formerly the Museum of Television and Radio.

At Yankee Stadium, Ken Derry and Tony Morante were extremely helpful, as were Rozaan Boone and Inbal Haimovich at Co-op City. For assistance with research about the Art Deco legacy of the West Bronx, I'm grateful to Andrew Capitan, Michael Kinerk, Glen Leiner, and Tony Robbins. For memories of Loew's Paradise, I benefited hugely from the recollections of Gerald McQueen and the expertise of Rebecca Shanor and Ross Melnick.

Others whose research suggestions, personal stories, and general insights about the Bronx helped in countless ways include Andre Aciman, Alan Adelson, Michael Agovino, Jacob M. Appel, Rick Bell, Leonard and Leo Benardo, Nancy Biderman, Charles Bleiberg, Harold Bloom, Michael

Bongiovi, Ray Bromley, Raphael Carman, Mary Childers, Barbara Cohen, Ira Cohen, Judith Crist, Laura Shaine Cunningham, Suzanne Davis, E.L. Doctorow, Christopher Rhoades Dykema, Nora Eisenberg, Mary Engelhardt, Gil Fagiani, Jules Feiffer, Arline Friedman, Mindy Thompson Fullilove, Ben Gibberd, Harrison J. Goldin, Vivian Gornick, Roberta Brandes Gratz, Helen Green, the late David Halberstam, the late Kitty Carlisle Hart, Pablo Helguera, Tony Hiss, Kenneth T. Jackson, Martin Jackson, Adrian Nicole LeBlanc, Randie Levine-Miller, Phillip Lopate, Elizabeth Macdonald, Francis Morrone, Rhona Nack, Regina Peruggi, Roberta Peters, Richard Plunz, Nick Raptis and his wife, the late Betty Kanganis, Morton Reichek, Patricia Twomey Ryan, Steven Samtur, Ellen Samuels, Joyce Sanders, Harry Schwartz, Helen Schwartz, Richard Sherwin, Edward Sorel, Maria Terrone, Eliot Wagner, Mike Wallace, Maureen Waters, Max Wilson, Carole Zimmer, and Destra Zabolotney.

For help with images, I'm extremely grateful to Carl Rosenstein, who made available his gorgeous photographs of Art Deco buildings of the Bronx; I'd also like to thank Phyllis Collazo, Nakyung Han, Maura Foley, and Natasha Perkel at the *New York Times*.

Many current and former *Times* colleagues provided insights, research tips, rich Bronx memories, and general encouragement over the long span of this project, among them Joseph Berger, Paul Goldberger, Barbara Graustark, Bernard Gwertzman, Michael Leahy, Sam Roberts, Marvin Siegel, and Dinitia Smith, along with all my friends and colleagues, past and recent, at the newspaper's City section, my longtime professional home at the paper.

Among the friends who offered moral support and sent their Bronx acquaintances my way, I'd like to thank Jonnet Abeles, Peggy Anderson, Peter Freiberg, Susan Hodara, the late Carol Horner, Caryn James, Patricia Kavanagh, Ann Kolson, Lawrie Mifflin, Ellen Pall, Carol Rocamora, Beverly Solochek, Andrea Stevens, and Barbara Strauch.

I'm grateful to my agent, Mary Evans, to David McBride, formerly of Routledge, now at Oxford University Press, for early support of this project, and especially to Eric Zinner at New York University Press for helping make the book a reality and for his astute editorial guidance all along the way. Everyone at NYU Press was immensely helpful; I'm particularly grateful to Ciara McLaughlin, Eric's assistant; Despina Papazoglou Gimbel, the press's ace managing editor; Fredric Nachbaur, Betsy Steve, Joe Gallagher, and Brandon Kelley in the press's marketing and publicity department; and Liz Cosgrove, who designed the beautiful cover.

Many years ago, when I first began thinking about a book on the Grand Concourse, I got support from Penn Kimball, a professor and later a colleague at Columbia Journalism School. The late William Ewald of the *New York Daily News* urged me to pursue the project when a book about the Bronx was about the last thing anyone wanted to publish. Gene Roberts has offered encouragement ever since I first went to work for him a quarter of a century ago.

The lion's share of gratitude goes to my husband, Andy, and my daughter, Sarah. Simply put, this book exists in large part due to their love, understanding, and most of all enormous patience during the long span of this project.

Introduction

HERE ARE TWO SCENES of life in the middle of the last century on and near the Grand Concourse in the borough of the Bronx in the city of New York.

≡ • ≡

One scene takes place a few years after the Second World War in a tiny candy store called Philly's, on Sheridan Avenue near 165th Street, just east of the broad, tree-lined boulevard that cut a majestic north-south swath through the borough. It is a September afternoon, and the place is jammed. Children are lined up along the lunch counter and bunched together near the candy counter, agonizing over what to choose among an array of liberty streamers, twizzlers, and malted milk balls. A trio of housewives, laden with shopping bags from Alexander's and Loehmann's, the emporiums up on Fordham Road, stop by to grab a quick cup of coffee before heading home, most likely to an apartment with a sunken living room and wraparound windows in one of the smart Art Deco buildings that line the neighborhood's main thoroughfare.

With Yom Kippur just a week away, some of these shopping bags contain the black suede pumps and mid-calf-length dresses of crackly black faille being featured in the better department stores, perfect for High Holy Day services at one of the many local synagogues. A couple of women are probably comparing notes on *The Next Voice You Hear,* a tear-jerker starring James Whitmore that is playing at the palatial Loew's Paradise, the Bronx's four-thousand-seat re-creation of an Italian Baroque garden south of Fordham Road. Other conversations might strike a more anxious tone. The Korean War is raging half a world away, and though the generals predict that the enemy will soon fold, the Fordham University ROTC has just added 177 recruits to its ranks.

As people familiar with the rhythms of Philly's will recall decades later, a balding bachelor named Mersch generally occupies a perch at the far

Rita and Phil Barish, mainstays of Philly's candy store just off the Grand Concourse on Sheridan Avenue. At left, the man everyone called Mersch, a longtime Philly's regular. (Private collection)

end of the lunch counter, one ear cocked to catch the rising and falling din of the ball game on the radio. Down the hill, the neighborhood's sainted Yankees are battling the Detroit Tigers for the American League championship, and local hopes are riding high, though with the nation's premier baseball team literally in their backyard, few people can remember a time when they aren't. On the next seat down, a child named Richie Sherwin cracks ferocious bubbles with his gum and pores over his baseball cards, occasionally lifting his head to demand, "Hey, Mersch, you gonna take us to the game next week?" In the phone booth, a guy is making book.

Behind the counter, Rita Barish, the owner's wife, feverishly dishes out small blocks of halvah and whips up an endless procession of eggs creams, malteds, and ice cream sundaes; by day's end, the musty store will be suffused with the intoxicating aroma of hot fudge, and Rita's cheeks will be flushed with perspiration. As she works, her husband, Phil, futilely tries to keep the smaller boys from blowing soda straw wrappers across the store and the bigger boys from riffling through the girlie magazines, publications that will look tame to a later generation but strike the boys of 1950 as un-bearably racy. Phil is worrying, too, about the impromptu stickball game taking place on the patch of sidewalk outside the front door. "One day," he mutters to himself, "a ball's gonna come flying through the window, and boy, will somebody be sorry!"

In the evening, people from the neighborhood may stop by for a pint of Breyer's ice cream, or a frappe to round off supper, while the rest of the family listens to *Duffy's Tavern* on the radio or watches *The Goldbergs* on television, still something of a novelty in many homes. Molly Gold-berg, the memorable Bronx housewife with the thick Yiddish accent and dark hair piled atop her head, is a particular favorite. Though the character is purely imaginary, the creation of a savvy writer and producer named Gertrude Berg, the lulling texture of her days, not to mention her ficti-tious address—1038 East Tremont Avenue—strikes a powerful chord with her West Bronx listeners. Still later, the regulars will come trooping into Philly's, men content to linger for an hour or more, sipping sodas in summer, hot chocolate in winter, and puffing thirty-five-cent Gold Label cigars until the night-owl *News* and the *Daily Mirror*—two cents in those days—hit the streets.

⇒ • ⇐

The other scene also takes place on a day in September, but fifteen years earlier and a world removed from the cozy little dramas enacted within the sweet-smelling walls of Philly's candy store. This scene was christened the Bronx Slave Market by Marvel Cooke, a black investigative reporter who went undercover to describe its operations for the *Crisis*, a publica-tion of the National Association for the Advancement of Colored People and one of the few that speaks directly to a minority audience. The street-corner job mart, in nearly every respect a ghastly mirror image of a tradi-tional employment bureau, operates at several locations, one of which sits at 167th Street, just a few blocks west of the Grand Concourse. "Rain or shine, hot or cold, you will find them there," Cooke subsequently wrote of the scene she witnessed this late-summer day,

> Negro women, old and young—sometimes bedraggled, sometimes neatly dressed—but with the invariable paper bundle, waiting expec-tantly for Bronx housewives to buy their strength and energy for an hour, two hours, or even for a day at the munificent rate of fifteen, twenty, twenty-five, or, if luck be with them, thirty cents an hour. . . .

She who is fortunate (?) enough to please Mrs. Simon Legree's scrutinizing eye is led away to perform hours of multifarious household drudgeries. Under a rigid watch, she is permitted to scrub floors on her bended knees, to hang precariously from window sills, cleaning window after window, or to strain and sweat over steaming tubs of heavy blankets, spreads and furniture covers. Fortunate indeed is she who gets the full hourly rate promised. Often her day's slavery is rewarded with a single dollar bill or whatever her unscrupulous employer pleases to pay. More often, the clock is set back for an hour or more. Too often she is sent away without any pay at all.

Among the regulars is an outspoken young woman named Millie Jones, and the windows that wrapped around the corners of the Art Deco apartment houses on the Grand Concourse are particular instruments of torture. "Mrs. Eisenstein had a six-room apartment lighted by fifteen windows," Millie Jones tells Cooke. "Each and every week, believe it or not, I had to wash every one of those windows. If that old hag found as much as the teeniest speck on any one of 'em, she'd make me do it over."

The exigencies of observant Jewish life prove no more forgiving. The young woman adds,

> Say, did you ever wash dishes for an Orthodox Jewish family? Well, you've never really washed dishes, then. You know, they use a different dishcloth for everything they cook. For instance, they have one for "milk" pots in which dairy dishes are cooked, another for glasses, another for vegetable pots, another for meat pots, and so on. My memory wasn't very good, and I was always getting the darn things mixed up. I used to make Mrs. Eisenstein just as mad. But I was the one who suffered. She would get other cloths and make me do the dishes all over again.

Edith Gumbs, a small, sturdily built woman who was born in the British Virgin Islands, did not arrive in New York City until the late 1930s, and so was not present the day Cooke showed up in the Bronx with her reporter's notebook. Still, Edith Gumbs knows this face of the Grand Concourse well. Every day for years, she has been leaving her apartment in Harlem at five in the morning and walking fifty blocks up to the Bronx—at the end of the day she will retrace her footsteps—where she earns five dollars for eight hours of scrubbing the parquet floors and scouring the pots and pans of the boulevard's doctors, judges, furriers, and other well-heeled residents. To make the strongest first impression when being given the once-over by prospective employers, she wears a nice coat and her best hat.

By the time the Gumbses move to Morrisania, a working-class community east of the Grand Concourse, Edith's son, Bob, has already started collecting his own notions about the street where his mother spent so many wearying hours. The lessons of de facto segregation as practiced in the North are being indelibly etched in his mind. When the boys he hangs out with so much as set foot on the boulevard, the police promptly shoo them away, sometimes reaching for their guns even before the boys have turned and fled. The day Bob Gumbs's uncle came to the neighborhood in search of an apartment, he was reminded in no uncertain that the street was reserved for whites only. "And on Sundays," Gumbs recalled half a century later, "when we used to take walks to the Grand Concourse, the only black person I ever saw was a super coming out of a basement with a garbage can over his head."

$$\Longrightarrow \quad \bullet \quad \Longleftarrow$$

The Grand Boulevard and Concourse, the four-and-a-half-mile-long, 182-foot-wide thoroughfare completed in 1909 and built originally to accommodate fast horses and horse-drawn carriages, was from the first far more than just another street. An engineering marvel nearly two decades in the making, the Grand Concourse was the creation of a brilliant and almost dreamily idealistic Frenchman whose greatest American achievement was strongly reminiscent of the famed Champs Elysees in Paris, a detail that generations of residents invariably mentioned when extolling the place where they lived.

The boulevard was born at a moment when the borough's population and prestige were poised to explode, a moment of all but tangible optimism and seemingly unlimited possibilities. Almost from the first, the street had a mythical significance. For huge numbers of upwardly mobile Jews, and from the early 1920s through the late 1950s, the Grand Concourse represented the ultimate in upward mobility and was the crucible that helped transform hundreds of thousands of first- and second-generation Americans—mostly Jewish but also Irish and Italian, along with smatterings of other nationalities—from greenhorns into solid middle-class Americans. Immigration was remaking the city and the nation in profound ways during those years, and it was to the Grand Concourse and similar destinations that countless immigrants aspired. The newcomers themselves rarely made their way to the boulevard proper. But with increasing force and velocity, the generations that came after them escaped the congestion and squalor of the Lower East Side, heading first to the tenement apartments of the East Bronx and finally to the smart and spacious precincts of the West Bronx.

As was the case with other important New York City thoroughfares—West End Avenue in Manhattan is a notable example—the very name Grand Concourse was synonymous with the idea of making it in America. Over and over, the American Dream was reenacted along its broad flanks, so often that the words *Grand Concourse* came to embrace not just a street but a way of life. It was no accident that novelists such as Avery Corman and E.L. Doctorow, two notable sons of the West Bronx, portrayed both the street and its world so eloquently in their fiction and that celebrated Bronxites ranging from director Stanley Kubrick to pop singer Eydie Gorme were forever linked to the geography of their childhoods. "My dear, it was Paradise!" the film critic Judith Crist once said a bit breathlessly in a conversation about coming of age in the neighborhood. She was speaking about the borough's celebrated movie theater, of course, but in a very real sense her words embraced the whole of her West Bronx girlhood.

The Concourse, as it was affectionately known, represented for many people a true street of dreams, an unblemished symbol of having unequivocally arrived. Those who could not make it to the street itself lived proudly nearby in its reflected glory, sometimes literally in its shadow, because the boulevard sat atop a steep ridge, and the sleek Art Deco apartment houses along its edges towered over the lower-lying structures on the side streets.

Although the planning and architectural accomplishments that defined this part of the city often received short shrift by virtue of their location far from Manhattan, these achievements were nonetheless considerable. With their sunken living rooms, airy corner windows, and stylishly attired doormen, the Art Deco buildings of the West Bronx represented the ultimate in both urbanity and modernity. Nor were they the neighborhood's only jewels. At the famed Concourse Plaza Hotel, on the boulevard at 161st Street, the powerful Bronx Democratic organization feted senators and presidents, and the best bar mitzvahs and wedding receptions were held in lavish ballrooms adorned with flocked crimson wallpaper and hung with crystal chandeliers. At Loew's Paradise, a populist palace erected at a moment when Hollywood reigned as the ultimate dream factory, real goldfish swam in a marble fountain, and décor included ten thousand dollars' worth of artificial birds and a ceiling ablaze with twinkling stars and drifting clouds.

At Temple Adath Israel, at the corner of 169th Street, one of the majestic Jewish houses of worship along the Grand Concourse, prayers were sung for a time by a silvery-voiced young cantor named Richard Tucker, who became a leading operatic tenor of his era. Fordham Road, which bisected the Grand Concourse midway, was the Bronx's answer to Fifth Avenue. At 161st Street, just steps to the west of the boulevard and so close that you could hear the cheering when Babe Ruth or DiMaggio hit a home run, sat the fabled Yankee Stadium, home of the powerhouse team whose members roamed the neighborhood like gods.

Like many memory-glazed places, this one had its nightmare elements. Although it is easy to romanticize the life lived on and near the boulevard, that life had a darker side. It could suffocate as powerfully as it could nurture. The passage to adulthood, never uncomplicated, was no less treacherous for a young person living in a handsome apartment building on a prestigious street. For every garment-factory owner who basked in the attainment of his bourgeois dreams, a budding poet was suffocated by the relentlessly middle-class values that engulfed him. The emphasis on material success that pervaded this world left little room for the eccentric, the nonconformist, the late bloomer, or the overly sensitive. "If you admired a sunset," said the Bronx-born cartoonist and writer Jules Feiffer, who escaped the borough of his birth as fast as he could, "everyone assumed you were gay." If you wanted to write sonnets rather than practice medicine, you were probably in the wrong place. Many people wouldn't have lived there for the world. Others spent a lifetime trying to escape.

⇒ • ⇐

For nearly half a century, the main identity of the Grand Concourse and its neighborhood was that of a mecca for immigrants and their offspring, and in this respect the area's fortunes are a key strand in the story of the American city. But the boulevard and its surroundings encapsulate that story in another way. One day they were intact, and seemingly the next they collapsed, shattered like fragile crystal; at least that was the traditional narrative, one that only decades later was examined in more critical and nuanced fashion. Beginning in the early 1960s, and almost overnight, or so many longtime residents felt, the West Bronx was transformed from a stable middle-class community into a tumultuous, frightening place inhabited largely by poor and sometimes dysfunctional newcomers. Crime and jagged-edged social disorder, much of it born of drug addiction, raked the streets and their inhabitants—both old-timers and newcomers—like splinters of broken glass.

The litany of troubles is in many respects sadly familiar: Because of a complex web of deep-seated social, economic, and political forces that were operating throughout New York City and echoing far beyond, low-income blacks and Puerto Ricans from the South Bronx began moving north and west into neighborhoods that had previously been socially cohesive, economically sturdy, and, most significant in the eyes of many who lived in those areas, almost entirely white. As the challenge of responding to these changes intensified, municipal indifference seeped into deliberate neglect and combined with dubious and often criminal real estate practices to rip apart an already fraying social fabric.

Steps away from the Art Deco beauties along the Grand Concourse, arson became so commonplace that the night sky seemed perpetually ablaze. The increase in disarray and violence—in an area once largely free of both— gave birth to a never-quite-disguised racism that in many eyes had been lurking just below the surface. Anxiety tumbled into panic. If one apartment was burglarized, the tale was repeated so often that a hundred apartments seemed to have been ransacked. If one shopkeeper was held up at gunpoint, the stories swirling around the neighborhood suggested that every owner of a local business had been mugged. Truth blended with rumor, and the truth was bad enough: the Forty-fourth Police Precinct, which covered a neighborhood directly west of the Grand Concourse, at one point had the unhappy distinction of reporting more crime than any other precinct in the city.

In the late 1950s, when the Cross Bronx Expressway began slicing through the West Bronx, deeply rooted communities on either side of the Grand Concourse were literally ground underfoot. In 1968, when Co-op City opened the first of more than fifteen thousand apartments in the northeast Bronx, many residents felt as if a sluice gate had opened. Almost single-handedly, Co-op City appeared to drain the lifeblood from the Grand Concourse. Entire buildings seemed to move en masse.

Though the solid apartment houses along the boulevard largely escaped the fires that raged through the borough's southern portion, their populations turned over almost 100 percent. As out-migration increased, building after building on the side streets and the streets parallel to the Grand Concourse were abandoned, along with notable structures on the boulevard itself. Other storied institutions on the Grand Concourse slipped into decay. By the 1970s, much of the West Bronx was unrecognizable, even to those who remembered the place as it had been just a few years earlier. The entire southern half of the boulevard and the communities along it were slapped with the label "South Bronx"—a synonym for the bleakest, most hopeless urban devastation.

In its despair, the Grand Concourse was not alone. Harlem, Central Brooklyn, and parts of cities around the country experienced similar upheaval. Yet unique to the Grand Concourse and the adjacent neighborhoods was the speed with which convulsive change occurred—at least the speed with which longtime residents experienced that change. The transformation seemed to arrive so suddenly that it felt almost apocalyptic in its scope and impact.

On the heels of these traumatic changes came a desperate search for explanations. Was the culprit Co-op City, so often cited as the villain of the piece? Was it the Cross Bronx Expressway, as Robert Caro argued powerfully in his 1974 biography of Robert Moses, the visionary urban planner who built the highway so loathed in the borough it bisected? Was it articles in local newspapers about the changing racial makeup of the West Bronx and the consequent departure of so many longtime residents, articles that would be quoted verbatim four decades after they were published, then excoriated as having fueled the exodus? Was it that the West Bronx had been so unwelcoming to minorities for so many years, or simply that the largely hidden and long-unaddressed poverty and inequality embodied by institutions such as the Bronx Slave Market had finally convulsed an entire part of the city?

The search for explanations was ultimately futile, nor, in any case, could a single answer suffice to explain so great a transformation. The West Bronx was succumbing to the turbulence rocking urban areas around the country during those years, turbulence that sprung from complex forces and policies, many of which had been set in place decades earlier. Rising expectations among minorities warred with longstanding historical injustice. Newly built suburbs offered a seductive alternative to city dwellers, who were quick to abandon their walkups and row houses in response to the siren song of emerald lawns, backyard grills, and two-car garages. Generational changes also played a critical role in remaking places such as the West Bronx; the young always yearn to move up and out, and particularly in Jewish enclaves, which are famously unstable, the desire to break free of the old neighborhood was potent.

For a time, the story seemed to end there, in hopelessness and helplessness. But in yet another turn of the wheel, in a shift that has also echoed around the city and the nation, the Grand Concourse and the communities along its edges are being transformed once more. Starting in the 1980s, even in the midst of the greatest unrest, the boulevard and the adjacent streets began to reclaim some of their old energy and vitality, if not their lost grandeur, albeit in a very different accent—often literally. That transformation continues. Many apartments in the boulevard's handsome Art Deco buildings are home to working-class and sometimes middle-class African Americans, Latinos, and Asians, who describe their sunken living rooms and wraparound windows with the same glowing pride as the Jewish schoolteachers and garment-factory owners who preceded them half a century earlier. For the first time in decades, white families are also making their home on the boulevard. Apartments along the Grand Concourse are growing increasingly desirable—and increasingly pricy.

Members of this new generation of strivers, even some who arrived during the street's most troubled years, use almost exactly the same words as their predecessors to articulate their hopes and aspirations, and to express how living in a place that feels important fuels their dreams. Crime in the West Bronx has fallen dramatically, as it has throughout New York City, and drugs have loosened their grip on the streets. Though nearly a third of the borough still lives below the poverty line, though the world of the Grand Concourse is not what it was and the much ballyhooed rebirth of the Bronx is often more wishful thinking than reality, many parts of the West Bronx pulse with life.

⇒ • ⇐

For generations, the story of the Grand Concourse has been told largely from the perspective of the upwardly mobile, predominantly Jewish New Yorkers who settled the neighborhood between the 1920s and the 1950s. This is the audience whose collective heart beats a little faster at the evocation of all those egg creams and stickball games, the audience that remembers the best of midcentury Bronx—the best of midcentury New York, for that matter—and gained strength from coming of age in a cohesive, nurturing, and richly textured environment. If any group on the planet would like to turn back the clock, it is the children raised in the West Bronx in the mid-twentieth century. It is no wonder that the Bronx nostalgia industry is booming and that members of the Bronx Diaspora whose personal ads pepper the Web sites and publications that feed this nostalgia seem so hungry to reestablish even tenuous connections to their youth.

But the Grand Concourse did not exist in isolation. For uncounted blacks and Puerto Ricans, the boulevard was a street on which they almost literally did not dare set foot. Like Bob Gumbs's uncle, they could never dream of renting an apartment along the Grand Concourse until those apartments were worthless. Even once they arrived, many of the newcomers were driven there not by choice but by necessity, propelled by indistinct forces that were largely beyond their control. Far more disproportionately than earlier residents, they suffered from the crime that ravaged the neighborhood, even if they were sometimes also its perpetrators. If longtime inhabitants saw their living conditions decline, the newcomers never knew anything other than decline. They were its primary victims, and unlike their predecessors, they had little means of escape.

This book tells the story of the Grand Concourse and the communities along its flanks, a world of grandeur followed by desolation followed by burgeoning optimism, a strivers' row laid waste by the hidden poverty and inequality that existed not far from its borders, a place that despite its tumultuous history remains a powerful symbol of the borough's identity even as it struggles to find its footing in a new century.

Though there is no street exactly like the Grand Concourse, the changes it experienced were hardly unique. These experiences were shared by many U.S. cities during those years, and for that reason the arc traveled by the boulevard and its neighboring communities demonstrates in lapidary detail how neighborhoods rise, fall, and struggle to be reborn. In part because change washed over the Grand Concourse so quickly and dramatically, the boulevard emerged as one of the most potent symbols of urban disintegration. Like a specimen in a bell jar, the story of the Grand Concourse provides a textbook example of how the American city fared in the last half of the twentieth century.

But the Grand Concourse was a unique place, as I discovered when I began exploring the history of the boulevard many years ago. Time after time, people I interviewed asked me if I wanted to tell the story of the Grand Concourse because I had grown up in the neighborhood. I hadn't, I replied, but I wished that I had. This realm held people in a peculiar thrall, even those barred from its precincts. When people talk about the Grand Concourse, an unexpected note often echoes in their voices. The memories are not necessarily sweet; especially for those denied access to this world, they are sometimes just the opposite. But invariably the recollections are vivid and intense, and people are extraordinarily eager to convey what it meant to be part of this place, or even to be excluded. For good and for ill, this street and all it represented etched lives in deep and indelible ways.

PART I

A Promenade for the Bronx

1 ⇐

"A Drive of Extraordinary Delightfulness"

IN THE LAST HALF of the nineteenth century, the sparsely populated acres blanketing southern Westchester County—the territory that would one day be known as the Bronx—might have struck most New Yorkers as the dullest place on earth; the action in those years occurred largely in Manhattan. But to a quartet of young municipal engineers who spent their workdays mapping the Town of Morrisania in the early 1870s, at least one portion of this area was rich with attractions: the expanse called Bathgate Woods that was thick with old-growth forest and vegetation. Bathgate Woods, later known as Crotona Park, was among the few remaining tracts of local wilderness and was so beautiful that in 1883 a city official described its trees as the prettiest of any south of the Adirondack Mountains.

During the week, the four civil servants busied themselves with compass and tripod, trying to impose order on the crazy-quilt street system north of the Harlem River. Their weekends, however, were their own, a time when these youths, barely out of their teens, could release some high spirits in one of the undeveloped tracts of land that were fast disappearing from the scene.

Bird's-eye views of geographical features flourished in the middle and late nineteenth century—one of the most famous images of the Grand Concourse is presented in such a fashion—and a bird's-eye view of what eventually became the Bronx would have depicted an area with two very different personalities. The irregularly shaped forty-two-square-mile expanse—the only portion of what would become New York City that was attached to the mainland—was bordered on the west by the Hudson and Harlem rivers and on the east by the East River and Long Island Sound. The portion to the east of the Bronx River was largely flat and in some places marshy, whereas the section to the west was so hilly that when streets were built, many had to be equipped with steps to march them up the steep slopes. The most noticeable features of the terrain were three massive wooded ridges that ran from north to south; the westernmost ridge overlooked the Hudson and Harlem rivers and the New Jersey Palisades; the easternmost ridge passed through Bathgate Woods.

Bathgate Woods was the destination of the four adventurers. Every Saturday afternoon during the hunting season, they met at their office at 138th Street and Third Avenue, then headed north aboard the Huckleberry Horsecar, so named because its leisurely pace gave passengers a chance to scramble off and pick berries along the way. At the final stop, they shouldered their rifles and trudged up the hill to an old stone hunting lodge perched near the crest of the ridge. The innkeeper was an aging Frenchman, and his manners and accent would have offered a welcome touch of home to one of the four visitors, a young man named Louis Risse who had come to America just a few years earlier from a small village in Alsace-Lorraine.

At four the next morning, the quartet set off in search of the grouse, pheasant, and quail that roamed the massive central ridge. Through acreage punctuated by jagged cliffs and gashed by rugged valleys, they ranged north to what was to become Van Cortlandt Park and the Westchester County line. From there, they headed south to what was left of the Morris estate, the once vast holdings of the family of the long-dead Lewis Morris, a signer of the Declaration of Independence.

Even for energetic young men used to tramping the woods as part of their jobs, the terrain must have posed a challenge. But to Risse, the newly transplanted Frenchman, those rocky expanses also offered the inspiration of a lifetime. The crisp autumn days, and in particular that great ridge, lingered in his mind. Perhaps it appeared in his dreams; it certainly haunted his waking hours. Two decades later, that ridge inspired him to create a brilliantly designed street—a Grand Boulevard and Concourse, he called it—that ran almost the entire length of the borough that was his home, profoundly shaped its development, and exerted a powerful ripple effect that extended far beyond its borders. In 1902, a moment when the project was mired in political infighting and still seven years away from completion, Risse published a brief but moving account of its conception that opened with a vivid description of those excursions in the woods.

⇒ • ⇐

The part of the city where those weekend expeditions took place had changed a great deal since the first European settler, a Swede named Jonas Bronck, had arrived in 1639 and two years later purchased five hundred acres of American soil east of the Harlem River. From Colonial days on,

Surveyors at work in the Annexed District, also known as the Twenty-third and Twenty-fourth Wards. That part of southern Westchester County eventually became the West Bronx. (*The Great North Side*)

the territory had been part of Westchester County, and by the mid-nineteenth century, the landscape was dotted with farms that produced fruits, vegetables, meat, and dairy products, along with country estates, market villages, and fledgling commuter towns—all signs of New York City's relentless push uptown. Epidemics of cholera and yellow fever also helped drive population north; these ailments were often vaguely attributed to "bad air," and bracing country breezes seemed the ideal antidote.

The first villages had arrived hard upon the construction in 1841 of the New York and Harlem River Railroad. Wherever a station was built, there a community sprouted: Melrose, Mott Haven, Fordham, Tremont. The earliest settlers of Morrisania, for example, were skilled workmen from the Bowery who in 1848 followed the rail line north, partly to be within commuting distance of their jobs in Manhattan and partly in search of an environment where their children could grow up close to nature. Within a decade after their arrival, Morrisania boasted a hodgepodge of one-family and two-family wooden houses along with a blacksmith and a few shops, the spire of St. Augustine's Roman Catholic Church providing the only hint of skyline. The oldest known photograph of the Bronx, an 1856 image of the village of Tremont—a community that sat along the route the Cross Bronx Expressway would follow a century later—shows barns, stables, houses, taverns, and a sprinkling of other local businesses, with cultivated fields in the distance.

Other settlements were far more sumptuous. As summering outside the city became an increasingly chic practice, well-heeled New Yorkers built lavish estates on these undeveloped acres and erected imposing villas, many of them generously gabled affairs topped by chimneys and ringed with porches. Some families journeyed to their estates entirely by carriage, driving north along Central Avenue, the favorite route to the fashionable Jerome Park racetrack and a thoroughfare that eventually took the racetrack's name.

By 1860, only about seventeen thousand people lived in the area. But the year 1874 brought a watershed event, the annexation to New York of a huge swath of southern Westchester County—the towns of Kingsbridge, West Farms, and Morrisania. On the stroke of midnight, nearly a score of villages dotting a twenty-square-mile area—everything south of Yonkers and west of the Bronx River—were incorporated into the city. Some twenty-eight thousand new New Yorkers were created in an instant.

The acquisition of the so-called Annexed District—the western half of what is now the borough of the Bronx—marked the first expansion of the city outside Manhattan. The newly acquired land also provided a canvas on which the young civil servant from Alsace-Lorraine left his remarkable legacy.

$$\Rightarrow \quad \bullet \quad \Leftarrow$$

Louis Aloys Risse entered the world in March 1850 in the town of St. Avold, near the German border. Family legend had it that he came from impressive stock, and in an account of his birth written nearly half a century later, he described his mother's family as notable for "conspicuous ancestry." His maternal grandfather had been a decorated officer in Napoleon's army, and he himself was born in a former castle that was once home to a duke.

Though nudged toward a career in the priesthood, Risse quickly rejected the idea. Instead, he headed for Paris, where he studied drawing and painting, and at the age of seventeen, he sailed for America, envisioning the visit as the first stop on a round-the-world journey. But he got no

Louis Risse, the visionary engineer who designed the Grand Concourse, in his office in the late 1800s. Though some people dismissed him as "the crazy Frenchman," the boulevard he created came to rank as one of the premier streets in his adopted country. (Museum of the City of New York)

farther than the country acres north of Manhattan. Risse apparently fell in love with the place, so much so that he decided to make his permanent home there and pursue a career as a civil engineer. He remained in the Bronx for the rest of his life, the last thirty-six years in a mansion on what was known as Mott Avenue, a street that would be incorporated into the Grand Concourse around the time of his death.

Risse could not speak a word of English when he arrived in the United States in 1868. Yet thanks to his professional training and his own capabilities, he quickly landed one engineering job after another, each more prestigious than the last. After working briefly as a railroad surveyor, he completed a street map of the town of Morrisania and helped draft a survey map of what became the Annexed District, officially known as the Twenty-third and Twenty-fourth Wards. There followed a series of municipal posts: assistant engineer in the Park Department, superintendent of roads in the Annexed District, and, in 1891, chief engineer in the district's Department of Street Improvements.

Risse's old hunting ground, the terrain he knew so intimately, ran like a golden thread through those years. At the Park Department, he personally surveyed the territory. As superintendent of the Annexed District, he was responsible for maintaining existing roads and charting the route of new ones. Over nearly three decades, he devoted himself to the taming and ordering of these acres, a task he mastered with growing proficiency.

Risse's career progressed so smoothly and speedily, it is impossible not to speculate on the qualities that helped a lowly civil servant with a polished continental manner but no apparent political connections rise so fast and accomplish so much. Certainly he was smart, ambitious, and energetic, an immigrant who eventually ascended to the post of chief mapmaker of the newly consolidated metropolis. A brief biographical sketch written during the height of his career describes him as "hale, hearty . . . of fine and pleasing presence . . . gentle, suave of manner." Such encomiums were typical of the age, but in Risse's case, they have the ring of truth. The subject of these generous words must also have been something of a poet; although some people dismissed him as "the crazy Frenchman," his writings are informed by an almost lyrical sense of what his adopted city might one day become.

In his vision of the New York of the future, particularly the arteries that would link its disparate parts, Risse focused not simply on getting from here to there but, more important, on the agreeableness of the journey—the ease, the charm, the logic. Like the best urban thinkers, he possessed a sweeping imagination and seemed able to visualize, often with almost prophetic accuracy, the face of the future. His was hardly a city for everyman; the New Yorkers on whose needs Risse focused were not the huddled masses but elegant boulevardiers who could have stepped out of a Manet painting. Nevertheless, many of the values embodied in his work had a high social and even moral purpose. And in an era in which the vast majority of the population lived pinched lives in cramped and dreary quarters, Risse believed passionately in the importance of creating beautiful open spaces and crafting efficient and attractive ways of traveling to and from them, even if a fine carriage was needed to make the journey.

At the moment that the first glimmerings of what would one day be the Grand Concourse entered Risse's mind, development north of the Harlem River was proceeding haphazardly and, in his considered opinion, quite stupidly. The Annexed District fell under the jurisdiction of what was officially known as the city's Department of Public Parks, and due largely to the political paralysis that resulted from a mayoral administration that changed hands every two years, the district seemed fated to sit unimproved for some time. Perceived as suburban rather than urban, the terrain struck many people as more suited to a parklike system of streets than a crisp, city-style layout. In addition, New Yorkers increasingly regarded the Bronx as Manhattan's playground, a recreation area to serve a crowded island whose population was growing by 6 percent every year.

Between 1884 and 1888, the city spent upward of nine million dollars to acquire nearly four thousand acres of raw parkland, most of it in the northern Bronx. A few years later, *King's Handbook of New York City* reported that the great expanses of green in the North Bronx—among them Van Cortlandt Park, Pelham Bay Park, and Bronx Park, future home of both the New York Botanical Garden, founded in 1895, and the New York Zoological Society, which opened its doors in 1899—had become immensely popular among people able to make their way to them. These tracts were, the guidebook reported, "already frequented by those who wish a rustic outing in the wild woods and pastures," used for picnics in summer and skating parties in winter. Yet visitors were few. Despite the enormous outlay of money for open space, no one had bothered to create a system of roadways linking the greenswards of the Bronx to the populated portions of Manhattan, and those who ventured north reached their destination only after a long and muddy journey. Thanks to municipal indifference, these lush areas sat in splendid isolation, far distant from the city's heart.

⇒ • ⇐

Toiling in and out of government during those years, Risse fretted quietly about this state of affairs, but a single individual, even a talented and hardworking public servant, was hardly in a position to rectify the situation. Many of his fellow Bronxites were not so patient. Increasingly fed up with the government's apparent inattention to local development, they began agitating for action, making their case so noisily and so persuasively that the state legislature finally stepped in to investigate the whole mess. A Department of Street Improvements for the Twenty-third and Twenty-fourth Wards was set up, and in 1890, Louis J. Heintz was elected as the agency's first commissioner.

The young scion of a wealthy family, Heintz arrived at the job with no previous experience in government. Born in 1861, he started out in his uncle's brewery, married the daughter of a millionaire brewer, then chucked the beer business for a career in public service. He held the commissioner's job for only two years—he died in March 1893 from pneumonia contracted during a trip to Washington to attend President Cleveland's second inauguration. Nevertheless, during his brief time in municipal government, Heintz made one prescient decision that transformed the face of the borough: he appointed Risse as his chief engineer.

It was a daunting assignment. Although the few roads that existed in the district were almost indescribably awful, the building of replacements had emerged as one of the most politically fraught issues of the day. In testimony before the state legislature in Albany, a prominent Bronx attorney and reformer named Matthew P. Breen painted a grim picture of the byways of the Twenty-third and Twenty-fourth Wards, describing them as "elongated mud ponds, punctured here and there with turbid pools of stagnant water and malodorous filth, the ever present parent of disease and death." As for the political shenanigans that accompanied the building of new roads, "A man might go to bed one night with a street apparently established in front of his house, and before he retired to rest the next night the plan would be changed, and he might find the street located in the back of his house."

Local pressure for development was not the only force helping to shape the Annexed District during those years. The city's prestigious Rider and Driver Club was clamoring for a speedway along which members of New York's Four Hundred could race their fast horses, and club president John De La Vergne, a prominent local businessman and a friend of Risse's, having made no headway with his scheme of establishing such a route in Central Park, was taken with the idea of converting Jerome Avenue into such a thoroughfare. Despite Risse's affection for De La Vergne, he was personally dubious, foreseeing a day, all too soon, when noisy and smelly commercial traffic would transform the avenue into the last place one would want to enjoy a brisk canter on a pleasant afternoon.

No one will ever know who first envisioned placing that speedway atop the ridge that bisected the borough, a site occupied at the time only by an ordinary street bordered by farmland and a smattering of Victorian houses. Some early historians credited Heintz with the initial inspiration and relegated Risse to the role of designer and executor of the plan, demoting the chief engineer of the Annexed District to little more than a high-class nuts-and-bolts man. But given the eloquence with which Risse described his vision for the street, and given all he accomplished later in his career, the far likelier scenario is that the age-old practice of underlings doing the work and bosses taking the credit was standard operating practice in turn-of-the-century Bronx and that the conception of the boulevard was his and his alone.

If such was the case, Risse would understandably have cared deeply about setting the record straight, at least the record as he saw it. In 1902, more than a decade after the earliest conversations about the Grand Concourse, Risse published a pamphlet insisting that he was the man who gave birth to the idea and the individual responsible for the inspired and politically astute yoking of two separate issues—the campaign for a speedway and the need to link the parks to the north with the population centers to the south.

Although physically a modest publication, just eleven small pages of text, the treatise in which Risse lays out this sequence of events is grandly titled *The True History of the Conception and Planning of the Grand Boulevard and Concourse in the Bronx* and stands as the major surviving document describing the intentions of the boulevard's creator. Recollecting the thoughts that passed through his mind after a conversation about the speedway with De La Vergne, Risse wrote, "it was then that I recalled that rocky ridge east of Jerome Avenue with which I had been so impressed during my hunting days. And at once the idea occurred to me that this ridge could well be utilized as a location upon which to build a broad and grand avenue that would serve the dual purpose of a connecting link between the Park systems and a Speedway or Grand Concourse." It is as if the proverbial light bulb had gone off in his head.

As Risse described that moment, the broad strokes of the plan came to him in a flash. For the Bronx, geography was destiny—its trio of ridges led to a century's worth of development along their north-south axes—but it was Risse who first recognized the potential of this particular undertaking. In his mind's eye, the acres northeast of the Harlem River presented an ideal stage set on which to build a new city, and the crest of the enormous ridge that stretched north to Van Cortlandt Park loomed as a natural future boulevard. The engineering challenges were formidable—the steepness of the ridge meant that cross streets would have to tunnel through dense bedrock to pass under the Grand Concourse—but nonetheless manageable.

A few days after that conversation with De La Vergne, Risse and his boss traveled to Jerome Avenue to view the site. "To me," the engineer wrote, "that day is memorable." As the two men stood in front of a roadhouse that offered a superb view of the ridge and the scattering of houses along its crest, Risse outlined his plan. According to the subordinate's account of the conversation, Heintz expressed initial skepticism, but his mood brightened when Risse moved to a subject of considerable interest to an ambitious public official, the enormous amount of revenue the project could generate as a stimulus to residential construction. If Heintz didn't utter precisely the words Risse put into his mouth—"Have a care that you don't get me in 'Dutch' with this project of yours for a Grand Boulevard and Concourse"—he nonetheless gave his aide leave to proceed.

⇒ • ⇐

Armed with an official go-ahead, Risse set to work making a topographic study of the area. He worked quietly, and he worked alone. One of the few people in whom he confided was De La Vergne, who shared his enthusiasm for the project and, more important, provided a generous dose of financing at a critical moment. Before details about the project were made public, Risse presented his plans and sketches at a private meeting of the politically influential Schnorer Club, an organization of local business and political leaders headquartered in Morrisania. At that meeting, despite the fact that not a nickel of city money was available even to print and distribute the material describing the project, the Rider and Driver Club agreed to raise five thousand dollars for that purpose. De La Vergne himself led the way by pledging one thousand dollars on the spot.

Bronxites who showed up to inspect the designs at a series of meetings in early 1893 were equally taken with Risse's vision for the area. Though Heintz's death cast a brief pall over the project, on April 1, just a few weeks after the funeral, Risse officially filed his plans for the boulevard with Heintz's temporary successor, along with an extraordinary letter. In this letter, which is quoted in *The True History*, Risse described the ambitious scope of his vision in a series of quick but indelible strokes.

"What is wanted," he wrote, "is a street system adapted to the city that is to come, and not to the infant, so to speak, which exists today." In twenty years, he predicted, up to a million people might be living in the area north of the Harlem River. "All this means a tremendous tide of travel to and from what are now the 'New Parks,' and a boulevard will be needed to accommodate the constantly swelling hordes of pleasure-seekers who will visit the parks in numbers."

Even though the city had spent a fortune to acquire the new parks, no decent roads led to them. Thoroughfares such as Jerome Avenue and Third Avenue did not fit the bill, he pointed out, because as commercial routes they were already being eyed for elevated lines or trolleys, and soon would be so congested that they could never function as proper boulevards.

And what was a boulevard? As Risse described it, in language that was for decades quoted to underscore the street's true nature,

> A boulevard is a promenade, a drive, an avenue of pleasure, everything in fact except a commercial thoroughfare. Carriages and trolley cars cannot run on the same avenue, and the endless procession of the family parties, enjoying the air, beaux and belles, the long array of children in charge of solicitious nurses and anxious mamas, and the other boulevard travelers, do not take kindly to trucks and freight traffic. Pleasure and business should be kept apart.

"Such a thoroughfare must be sufficiently elevated to give a commanding view in all directions," he added. As for this particular thoroughfare, which he originally christened the Speedway Concourse in a nod to its initial purpose, "nature has provided a grand boulevard for the North Side in the City of New York such as no other city in the world possesses. . . . The ridge, which is the line of the projected Boulevard, with respect to its beauty, has no parallel either in Europe or America."

Risse went on to spell out the specifics of the project: a boulevard 182 feet wide with a dirt-and-cinder bridle path down the center and separate roadways on either side for cyclists, pedestrians, and horse-drawn carriages, the sections separated from one another by tree-lined malls and from trolley traffic via a system of twenty-three underground cross streets. Then came the sales pitch. "No investments made by a municipality pay such large returns as improvements which attract visitors," Risse pointed out. In fact, he predicted a bit optimistically, visitors would be likelier to visit such a stunning and useful thoroughfare than even the parks to which it led.

An accompanying pen-and-ink drawing depicts a broad roadway lined with feathery trees that meet overhead to form a leafy arch. Splendid villas march along the edges of the street; jaunty horsemen in top hats trot down the central bridle path, flanked by carriages and strolling couples. The street stretches as far as the eye can see, its tiny inhabitants dwarfed by the vastness of the spaces surrounding them.

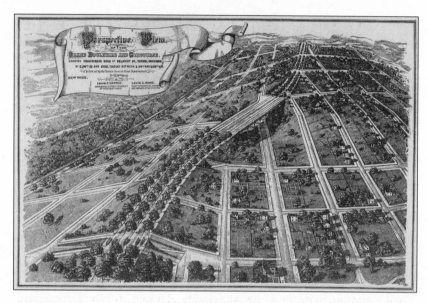

An early drawing of the Grand Concourse showing the landscaping that became a signature aspect of the boulevard's appeal. (*The Great North Side*)

The vibrancy of the scene Risse envisioned was breathtaking, all the more so given the current state of the land in question. Though by 1892 the borough's population was estimated at eighty-one thousand, a bird's-eye view of the route of the proposed boulevard would have revealed few signs of habitation. Notwithstanding the assortment of Victorian houses that lined that route, the road along which they sat was utterly unremarkable, barely mentioned in accounts of those years. The primary signs of life were, to the west, the bridges that spanned the Harlem River, the inns and roadhouses along Jerome Avenue, the Jerome Park racetrack and, to the east, the Fleetwood Park racetrack and the railroad. That was about it.

≡ • ≡

For the most part, the world of the civil engineer is wreathed in the most mundane of accoutrements—tripods, compasses, endless numbers of maps, and page after page of dry-as-dust calculations. But a set of remarkable photographs that show Risse and his staff at work during the years the Grand Concourse was taking shape put a deeply human face on their seemingly prosaic labors and offer an intimate glimpse of Risse's days as he labored over his most important project.

An 1892 drawing of the Speedway Concourse, as the boulevard was known early on. Although Louis Risse pictured villas along its flanks, history provided something with far greater potential to transform this part of the city: block after block of five-story and six-story apartment houses. (Museum of the City of New York)

The photographs were taken in Risse's office, about a mile to the southeast of where the boulevard was to begin. That office was housed in a tenement building at the corner of Third Avenue and 141st Street in Mott Haven—construction of the handsome Renaissance-style borough hall that opened in 1897 just north of Crotona Park lay a few years in the future. A calendar clock on the wall shows the date as March 22, 1894. Judging by the general messiness, the chief occupant of the office was not a particularly tidy individual, or maybe the nature of his work made it hard to be neat. Stacks of ledgers sit precariously on every available surface, and maps are pinned to the walls and spill wildly out of cubbyholes. Beneath the light cast by an overhead brass fixture, near a pigeon-hole desk piled with rubber stamps and other office equipment, a male secretary pecks decorously at a typewriter. At the center of the hubbub sits a grave-eyed man with a flowing mustache and bushy white hair, his somber black suit relieved only by the watch fob stretched over a rather broad stomach.

Working in such settings during the waning years of the nineteenth century, Risse and his associates produced the definitive images of the multilane, multilayered thoroughfare that would stretch some four and a half miles from 161st Street north to Mosholu Parkway, just south of Van Cortlandt Park. These images, in particular a pair of stunning maps dating from the summer and autumn of 1892, drawn with pen and ink on snowy

Louis Risse in front of the map for the project that was his greatest achievement, the four-and-a-half-mile-long, 182-foot-wide boulevard that eventually extended from 138th Street north to Mosholu Parkway. (*The Great North Side*)

linen, are almost as much works of art as the boulevard whose lineaments they trace. So meticulously crafted were these maps, they looked new more than a century later, and in their razor-sharp detail, they suggested the complexity and ambition of this project.

The larger of the two maps, a twelve-foot-long document drawn in August, traces the boulevard's leisurely, slightly sinuous path and includes a detailed inset depicting an elegant arrangement of two-way speedway, adjoining drives, and sidewalks, the six broad bands separated from one another by little fluff balls to indicate the rows of trees that would keep the arteries separate, civilized, and sweet-smelling. The map drawn a few months later, though more modest in size, almost literally radiates the optimism that wreathed the undertaking; the confident, florid signatures of Risse and Heintz, a vivid contrast to the painstakingly drawn little checkerboards that cover most of the rendering, are almost joyous. The day the two men signed their names to this document must have been an exhilarating one, marking the completion of one critical phase of the project and ushering in the activities that would transform the beautifully drawn image into reality.

As journalists became increasingly familiar with Risse's great project, their enthusiasm began to echo his. "The boulevard," James Barnes wrote in *Harper's* magazine in November 1897, "will have no equal anywhere in this country or in Europe. It follows the ridge carefully, and it would almost seem as if the latter had been prepared for its occupancy."

Even the early renderings make clear why the Grand Concourse was so often compared to the Champs Elysees, the sweeping tree-lined boulevard that was routinely described as the most beautiful street in Paris and perhaps the world. Both were glorious thoroughfares seemingly tailor-made for stylish boulevardiers, and Risse might well have been inspired by the fashionable street he knew so intimately from his student days in the French capital. But an equally powerful influence lay much closer to his adopted home.

In 1858, Risse's fellow urban visionary Frederick Law Olmsted and Olmsted's partner, Calvert Vaux, had won a design competition to create what would be the nation's first major urban park. The 840-acre expanse of green to be known as Central Park would be accentuated by rustic arbors, tree-framed vistas, and willow-fringed pools, all seemingly as nature had intended and in reality just the opposite. But what would have particularly interested Risse were the roadways. To protect pedestrians, a series of east-west arteries were routed under the two major north-south routes and buffered with greenery so that strollers, cyclists, and horseback riders could neither see, smell, nor be struck by speeding vehicles. The sophisticated circulation network represented a revolutionary achievement, although Risse, in integrating those underpasses into the design of a major urban street, expanded the concept considerably.

Around the same period, Olmsted and Vaux also created the two Brooklyn thoroughfares that despite some significant differences resembled the Grand Concourse in scope and spirit: Eastern Parkway, the two-and-a-half-mile street that stretched east from Grand Army Plaza, and the six-mile-long Ocean Parkway, which extended south from Prospect Park to Coney Island. And in an ironic twist of fate, had the history of the Bronx played out rather differently, Olmsted rather than Risse might have been the man who created its most important roadway.

In 1874, sixteen years after the start of construction of Central Park and about a decade after work had begun on the two Brooklyn parkways, Olmsted was drafted to prepare a master plan for the newly created Annexed District. By this point in his career, Olmsted had concluded that the Commissioners' Plan of 1811 that had parceled Manhattan into standard, two-hundred-foot-deep lots presented major drawbacks, one being that the grid could not be altered to reflect fluctuations in topography. Yet his proposal for the Bronx was not so different from Manhattan's checkerboards after all. "What is striking about the Bronx plan," writes Olmsted's biographer, Witold Rybczynski, "is how circumspect it is. There are no grand parkways or *ronds-points*." Olmsted had recently come to grief over a far more ambitious plan for the city of Tacoma, Washington, which the city fathers had rejected on the grounds that it was rather too imaginative, calling for blocks shaped like melons, pears, sweet potatoes, and in one alarming instance, a banana. The celebrated landscape architect had no intention of making the same mistake twice.

In reality, Olmsted's plan for the Bronx was subtler and more ingenious than it looked. The proposal, however, came with a high price tag, and for political reasons City Comptroller Andrew Haswell Green refused to approve the money needed to survey the property and buy the necessary rights-of-way. After more than a year's work, it was apparent to all that Olmsted would not be leaving his imprint on this part of the city.

≡ • ≡

Historians of the city have long debated the exact source of Risse's inspiration. But whatever or whoever put the configuration of the boulevard in his mind, one thing was clear: in every respect, the boulevard was the perfect road for its time, the quintessential reflection of its era. The street perfectly echoed the optimistic, can-do spirit that permeated New York City during the late nineteenth century, a spirit that expressed itself most dramatically in jaw-dropping engineering feats. The Brooklyn Bridge, yoking two great cities, opened in 1883. By the 1890s, the Manhattan skyline was soaring. The decade in which the very word *skyscraper* became popular saw the arrival of one after another tall building, each loftier than its predecessor: the twenty-one-story American Surety Building at 100 Broadway, the twenty-three-story American Track Society Building at Nassau and Spruce streets, the thirty-story Park Row Building.

In the Bronx, an electric trolley had replaced the old Huckleberry Horsecar, and an elevated rail line was pushing north along upper Third Avenue. Like the other ambitious undertakings transforming the iconography of the city in those years, the Grand Concourse embodied the prevailing belief that human hands and human muscle could accomplish just about anything.

All this frenetic bridge building and expansion of rapid transit reflected another hallmark of the period, the yearning to hammer together the disparate sections of the city into a single, integrated whole. The eastern part of the Bronx was annexed in 1895, and Greater New York was born three years later with the consolidation of the five boroughs into a single pulsing metropolis, "an eventful night of nights that sees a city born," as the banner headline in the *New York Journal* described the occasion that in an instant transformed New York into the world's second-largest city, trailing only London. In conceiving the Grand Concourse as a northern extension of Fifth Avenue, a thoroughfare that would shuttle the fashionable world of Manhattan to the rural expanses of the Bronx, Risse envisioned the street as an essential element of this unification, part of the glue that would bind two key units of the newly emerging city.

Some of the most winsome expressions of these twin themes of optimism and consolidation are contained in a little book called *The Great North Side, or Borough of the Bronx*, a collection of reports by officials and civic leaders that appeared on the eve of consolidation. This work, published in 1897 by the North Side Board of Trade, the Bronx's premier civic group, had the unabashed goal of "attract[ing] population, capital, and business enterprise to the borough." With few exceptions, surviving copies of *The Great North Side* have not aged well physically; the covers have turned a muddy green, and so fragile are the brown and brittle pages that they literally crumble to the touch. Yet even allowing for the overheated prose and the civic hyperbole characteristic of the age, to read the words of these turn-of-the-century movers and shakers is to sense the confidence that suffused the borough at this particular moment. These local power brokers clearly viewed the Bronx as an empty canvas on which the most extravagant dreams could be realized.

"We were never before better prepared to meet the exigencies of the situation than at present," wrote James Wells, president of the board. "We occupy a superb position geographically between the Hudson and the Sound. The section is renowned for its salubrity and magnificent scenery. . . . Our educational advantages are unsurpassed. Our public parks are among the most beautiful in the world." In a few years, predicted Louis F. Haffen, who had succeeded Heintz as commissioner for street improvements in the Twenty-third and Twenty-fourth Wards and who would go on to serve as the Bronx's first borough president, "our territory will not only be the garden spot of the metropolis, but will be a vigorous rival, so to speak, in commercial supremacy with that part of the city within the confines of Manhattan Island." The population of the Bronx, which half a century earlier at stood at only eight thousand, was climbing fast: twenty-

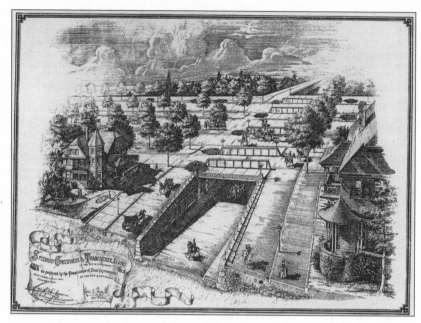

A drawing of the Grand Concourse, depicting one of a series of crosstown roads that would run beneath it. A rider on horseback emerges from the shadowy depths into the sunlight as if on the threshold of a new world. (*The Great North Side*)

four thousand in 1860, fifty-two thousand in 1880, and poised to reach two hundred thousand in 1900.

Risse's boulevard, though still in the distance, loomed as an integral rib of this Oz-like metropolis. "No time is fixed for the construction of this Grand Boulevard and Concourse," Haffen wrote, "but when completed it will be the most magnificent thoroughfare in the world."

Risse himself sounded hardly less effusive about the project. "It will be an improvement worthy of the great metropolis of the country," he predicted in the essay he contributed to *The Great North Side*. "The new Concourse will give a continuous sight of New York's beauties. It will be a drive of extraordinary delightfulness and practical convenience, and will offer the peculiar attractiveness arising from the sense that one may drive for miles without encountering an interruption in the road or a change in its character."

An accompanying series of etchings shows an almost futuristic-looking latticework of overpasses and underpasses stretching far into the distance. The roadway itself, fringed with tiny trees, slices through a checkerboard arrangement of blocks dotted with miniature houses. Here and there a doll-like figure wearing a long gown and carrying a little parasol surveys the landscape. A diminutive man on horseback emerges from the shadowy depth of one of the underground transverse roads, cantering into the sunlight as if on the threshold of a wondrous new world.

≡ • ≡

Wondrous, perhaps, and an excruciatingly long time arriving. Even given the delays and allegations of corruption that were a fixture of turn-of-the-century New York City, the Grand Concourse faced an unusually daunting series of setbacks. Initial enthusiasm curdled fast, replaced by volleys of recriminations, a cascading series of political shenanigans, and a few inconvenient acts of God.

Between that day in 1890 when Risse stood with his boss and gazed with him on the ridge where he used to hunt grouse and pheasant and the one in November 1909 when the boulevard was officially opened to traffic, nearly two decades passed. Not until 1895 did legislation pave the way for acquisition of the necessary property, and not until two years later, on August 28, 1897, did the city finally take title to the necessary land. Debate over which legislators' names would be listed on the bill authorizing the start of construction almost scuttled the project in its infancy. "The question," the *New York Times* remarked, "may involve the life of the entire twenty-million-dollar scheme."

The following year, a half-million-dollar appropriation ran into trouble in the hands of the new city's fiscally pressed government. "Of course the Concourse will have to be completed sometime," a state senator who opposed the allocation allowed peevishly. "The city has already accumulated three million dollars worth of land for the purpose, and it is out of the question for it to abandon the enterprise at this stage." However, the legislator added in an aside that must have chilled the hearts of the boulevard's supporters, "the work should not be undertaken just now."

The contract for construction was finally awarded in 1902. Borough President Haffen turned over the first shovelful of dirt on October 2 of that year, using a gold-plated shovel that was preserved for posterity. The Uvalde Asphalt Paving Company, the firm hired to build the project, swore that the job would be completed within one thousand working days, an estimate that proved off by 100 percent.

Part of the trouble involved an unexpected roadblock—a literal one—at 174th and 175th streets, where engineers came upon a deep valley that

One of hundreds of photographs taken by the city to record construction of the Grand Concourse between 1902 and 1909. They make the project look exciting, although living through seven years of digging and drilling must have taken its toll on residents of the Victorian houses along the street's route. (Bronx County Historical Society)

had to be bridged with a structure engineers called the "Great Wall of China" so a roadway could be built atop it. But many more of the obstacles were man-made, among them a scandal in which three commissioners for the boulevard were each paid eighty-five thousand dollars for what the *Times* described as "fat condemnation jobs." Before long, legislators were routinely describing the road as a giant boondoggle that would cost too much and, more to the point, "lead to nowhere." By the time the boulevard was completed, so much time had passed and the world had changed so profoundly that the side roads intended for horse-drawn carriages had to be paved to accommodate the newly fashionable automobile, not even a gleam in Risse's eye when he drew up his initial plans.

⇒ • ⇐

The ceremony marking the opening of the Grand Concourse took place on November 24, 1909, the day before Thanksgiving. From the point of view of the little band of prominent Bronxites who turned out for the event, the timing could not have been worse. A devastating winter storm swept through the Northeast that day, lashing New York City with special force. Driving sleet combined with thirty-mile-an-hour winds to cause multiple injuries and at least one death: a sixty-eight-year-old man who had climbed to the roof of his building on the Lower East Side to say his prayers was bashed in the head by the door of a coal scuttle, which fractured his skull. Elsewhere in the city, gusts hurled flower pots and window panes to the ground, sending chunks of glass and pottery onto passersby.

Up in the Bronx, at 161st Street, the official gateway to the new boulevard, some seventy-five men bundled in overcoats and mufflers stood shivering as acting Borough president John Murray presided over exercises that "were planned to be very simple originally," according to the *Times*, and "had to be curtailed on account of the discomfort entailed on the spectators by the bad weather." From the steps of the pedestal of the bronze statue honoring Louis Heintz, borough president-elect Cyrus Miller and his longtime predecessor Louis Haffen delivered short addresses and "speculated enthusiastically on the benefits the new road would bring to the Bronx." Afterward, the entire party was taken in automobiles along the length of the new roadway, which was to begin welcoming the public the following day.

Despite the cheerful rhetoric, remarkably little fanfare greeted a project nearly two decades in the making, and the sense of anticlimax was echoed in news accounts of the ceremonies. The brief article in the following day's *Times*, published beneath the headline "$1,000,000 Parkway Open: The Bronx Grand Concourse Starts Its Career under Stress of Weather," was buried deep inside the paper, nudged off the front page by reports of an arrest at the Metropolitan Opera House ("Sleuth in Evening Clothes Captures an Operagoer, Calling Him a Pickpocket"), the efforts of students to expand a local high school ("Girls' Witchery Won Eight-Story School"), and the travails of a Queens teenager who was discovered drunk in his classroom ("Bottle of Gin, Half Empty, Found in His Desk"). As was so often the case in the years that followed, an event hailed as transformative uptown seemed of singularly little interest to the reporters from Manhattan.

And, in fact, the price tag had been much higher than the headline suggested. Though estimates of the boulevard's cost fluctuated over the years, according to an article published in the *Times* about a decade later, the actual cost was $6.6 million—the equivalent of $114 million in 2008 dollars—a figure that included $3 million for acquisition of the necessary land and $2 million for expenses ranging from site preparation to the purchase of the shade trees that would be such a distinctive feature of the boulevard for generations to come.

The following Sunday brought what Bronx residents would have deemed a more appropriate response, a detailed article heralding the road's opening as a major moment for the borough and portraying its creators as among the great figures of the age. Those creators, the *Times* went on to say, were the late commissioner of street improvements, the man who had conceived the brilliant plan—"Idea Originated 20 Years Ago in Mind of Louis J. Heintz"—and Louis Haffen, his successor in that job and the man who made the road a reality.

In the case of Haffen, the accolade might have prompted a few raised eyebrows. Though in subsequent years, this devout Tammany operative was virtually written out of the history of the Bronx, he enjoyed an impressive political career, nourished by deep roots in the borough where he worked his entire life. Born in 1854, he never lived more than half a mile from his birthplace in the village of Melrose, and in 1897, the first time Bronx voters went to the polls to select a borough president, Haffen was their man. His constituents reelected him three times, allowing Haffen to preside over the Bronx during the years that its greatest street became a reality. But in August 1909, just six months before his term expired and in the wake of a two-year investigation, he was ousted from office by Governor Charles Evans Hughes, charged with "waste of public funds" and "political jobbery." Among the twenty-two counts of impropriety, at least one involved the construction of the Grand Concourse.

Haffen brushed aside the charges as the work of jealous political enemies. "This is a fine reward for twenty-six and a half years of honest, faithful, and efficient service to the people," he told a reporter indignantly. Many Bronxites apparently agreed. That autumn, when Haffen ran as an independent write-in candidate for borough president, he did surprisingly well, garnering half as many votes as Cyrus Miller, the eventual winner. Yet notwithstanding the praise heaped on Haffen that frigid day in November, this curious footnote to Bronx history could hardly have enjoyed sharing honors for one of the borough's most enduring achievements with the occupant of the job he deemed rightfully his.

Amid all the congratulatory back-patting, a single name was conspicuous by its absence, at least in news accounts of the occasion. The man whom history later christened "the father of the Concourse," the man who had described the boulevard in such evocative prose and detailed its lineaments in such elaborate maps and sketches, the man whose lacy signature was prominent on virtually every depiction of the road, was not mentioned once by the *Times*. Others compounded the omission; Risse's name did not appear, for example, in Stephen Jenkins's classic work, *The Story of the Bronx: 1639–1912,* which claimed that the idea for "the most magnificent boulevard in the world" had originated with Heintz as far back as 1890.

In addition to writing Risse out of the history of the boulevard's creation, the *Times* also skipped any reference to the messy battles that had swirled around the Heinrich Heine Fountain, which stood at the entrance to the boulevard. The fussy white marble concoction, known affectionately as the Lorelei fountain, depicted the legendary siren of Heine's most famous poem, "Die Lorelei," the Rhine maiden who lured sailors to their deaths with her irresistible singing and her equally irresistible beauty. In this incarnation, Lorelei perched atop a pedestal, surrounded by an assemblage of adoring mermaids and dolphins.

The work had been commissioned by the Empress Elizabeth of Austria as a present to the city of Düsseldorf, the poet's birthplace, and designed by Ernst Herter, one of Germany's leading sculptors. Upon its completion in 1893, Düsseldorf promptly rejected the gift, deeming a poet who was Jewish and wrote verse critical of German society unsuitable on several counts. A group of Americans of German descent then purchased the monument, the plan being to install it at the southeast corner of Central Park, in a neighborhood then home to many German Americans. Almost immediately, the agencies constituting the city's art police rejected this location on the grounds that so prominent a site was inappropriate for a creation honoring so partisan a figure—Heine being both German and Jewish—not to mention the fact that bare-breasted maidens seemed not the ideal thing to plunk down amid the staid dowagers who presided over the mansions along Fifth Avenue.

After six years of wrangling, the site on 161st Street was chosen, one advantage being that this neighborhood too was home to a vigorous German American community. The sculpture was unveiled on July 8, 1899, whereupon it was immediately defaced by vandals. Round-the-clock police were installed to keep an eye on the pretty Teutonic maiden lest anyone else tried to bother her, although as time would show, her troubles were far from over.

Although omitting any reference to either Risse or Lorelei, the *New York Times* did offer some unsettling reflections about the boulevard's future. Noting that no one had yet figured out how to connect its southern end with the streets of Manhattan, the newspaper predicted gloomily that "it may be several years yet before the Concourse will reach the stage where it can be said to fulfill all the functions it was originally designed for." And the paper sounded downright melancholy when contemplating the pace of residential construction along the Grand Concourse. In the years 1902 and 1903, builders had taken steps to erect only five houses and a single stable—at a total cost of just thirty-six thousand dollars—and the picture was hardly rosier the year the road opened. Fingers were crossed all around that activity would pick up once trains began rumbling on the Jerome Avenue subway line in 1917. "Brokers Expect Boom along New Roadway," the *Times* noted guardedly in a headline, "but Not Until Jerome Ave. Subway Comes."

Accompanying these disheartening observations was an uninspiring photograph of the transverse road running beneath the boulevard at Tremont Avenue. The picture didn't make the Grand Concourse look like a grand boulevard. It didn't make the street look like much of anything. If Risse had gazed at the photograph in the *Times* that Sunday morning, and it's hard to imagine that he didn't, the thought might have crossed his mind that he was lucky to have long since moved on to more promising pursuits.

⇒ • ⇐

After completing work on the Grand Concourse, Risse did something very much in keeping with the mood of the period. As the chief topographical engineer for the brand-new metropolis, a post he won the year of consolidation, Risse produced the first official map of Greater New York, a document that in some respects ranks in importance with the Commissioners' Plan of 1811. Although Risse's map did not carry the full force of law, it had two great virtues. The document was the first to show all the city's existing streets and, more significant, the first to offer a comprehensive and even grandiose plan for New York's future development, via rough layouts for streets and parks throughout the five boroughs.

"Unquestionably, the most interesting and imposing feature of the whole production is the tentative or proposed layout of an immense street system," Risse wrote in a report that accompanied the map. "This magnificent system, with its rectangular network of broad streets, diagonally intersecting boulevards, public squares and parks, canals, viaducts and bridges, and spreading areas divided up into large sections for residential and commercial purposes . . . is in its complexity of plan and vastness of extent without a precedent in the history of civilized society."

Not quite. Yet even allowing for the creator's hyperbole, the proposal had considerable merits. Risse predicted that New York City would someday be home to twenty million people. Here was a bold plan appropriate for such a vast metropolis, one that would extend the park system throughout the city, its parts connected via a series of parkways and boulevards much like the Grand Concourse. Along the waterfront, he proposed shops, recreational activities, and bridges to bind the boroughs even more tightly together.

Although many of Risse's ideas, particularly the ambitious network of superhighways and diagonal streets, never came to pass, the map itself was a dazzling piece of work. Risse had produced the original, inked and in full color, in just six months—incredibly quickly for a document of such scope and detail—and trimmed it with an elaborate border of nearly fifty pen-and-ink sketches of city views, based on his private collection of local scenes. Immediately after its completion on January 8, 1900, the map and its creator were shipped abroad so the document could be displayed in the United States section of the Exposition Universelle, the Parisian world's fair. There the map took first prize, and in a ceremony that must have been deeply emotional for a native Frenchman who had left his homeland more than three decades earlier, Risse was named an officer of the Legion of Honor.

The map's success inevitably prompted comparisons between Risse and another French engineer who played a critical role in shaping a U.S. city, Major Pierre Charles L'Enfant, whose plan for a gridiron street layout relieved by diagonal avenues and punctuated by public parks defined the look and feel of the nation's capital. But whereas L'Enfant always seemed the quintessential Frenchman, Risse's heart was in the Bronx, especially the mansion at 599 Mott Avenue, on the corner of 151st Street, that was his home for nearly half his life.

It was there, on a street that sat at the southern tip of the Grand Concourse and was eventually incorporated into the boulevard, that Risse and his wife raised three children and there that he retired in his fifties, long before the ribbon was cut on the project of his dreams. Visitors during those years were greeted by what his granddaughter, Marion Risse Morris, in an affectionate memoir, described as "a majestic figure in a pearl grey suit with a scarlet carnation at the lapel," a courtly individual who listened to French songs on the Victrola while poring over his collection of medallions and coins from around the world. It was in this house that Risse died at the age of seventy-four on March 10, 1925, a date marked for decades as a local anniversary. Fittingly, the founding members of the Bronx Board of Trade served as his pallbearers, and Risse's body was buried in Woodlawn Cemetery, near the crest of the ridge he loved.

2 ≡

"Get a New Resident for the Bronx"

THE NAME THEODORE DREISER conjures images of the windswept prairies of the Midwest or the grimmer precincts of turn-of-the-century Manhattan, the place some of his more hapless characters end up. Between 1904 and 1906, however, this writer so closely associated with the heartland lived in a drab apartment at 399 Mott Avenue in the Bronx, less than a mile south of the spot where great quantities of earth were being moved to make way for the roadbed of the Grand Concourse. Dreiser was miserable on Mott Avenue, and the experience seared itself so deeply into his consciousness that a few years after he moved away, he painted a harrowing picture of that chapter of his life in *The Genius*, his semiautobiographical novel about an ambitious young artist seeking glory in the big city.

The Genius, published in 1915, traces the struggles of a midwesterner named Eugene Witla, who, much like Dreiser himself, stumbles over one obstacle after another in his efforts to forge a career in New York. After suffering a nervous breakdown and working for a time as a laborer, Witla finally lands a job at a downtown advertising agency, and he and his wife settle into an apartment on Mott Avenue and 144th Street. But Witla is deeply ashamed to be living so far from the city's more prestigious quarters, and never more so than the evening when his boss, Daniel Summerfield, pays an impromptu visit to see his paintings.

> One night when [Summerfield] was riding uptown on the L road with Eugene, he decided because he was in a vagrom mood to accompany him home and see his pictures there. Eugene did not want this. He was chagrinned to be compelled to take him into their very little apartment, but there was apparently no way of escaping it. . . .
>
> "I don't like you to see this place," finally he said apologetically as they were going up the steps of the five-story apartment house. "We are going to get out of here pretty soon. I came here when I worked on the road."

Summerfield looked about at the poor neighborhood, the inlet of a canal some two blocks east where a series of black coal packets were and to the north where there was flat open country and a railroad yard.

"Why, that's all right," he said, in his direct, practical way. "It doesn't make any difference to me. It does to you, though. . . . You'd better move when you get a chance soon and surround yourself with clever people."

During the years Dreiser was describing, this section of the Bronx had become increasingly developed. One by one, stretches of farmland were making way for residential buildings, particularly the small apartment houses of the sort both Witla and his creator inhabited. The presence of the rail yards and the canal gave the borough's southern tip a gritty industrial air that did nothing to enhance its ambiance. At the time, the suggestion by Witla's boss that his ambitious employee get out as soon as possible seemed like excellent advice.

⇒ • ⇐

Dreiser, like his protagonist, didn't linger long in the Bronx. Although the squat apartment house where he lived survived into the next century, Dreiser left the borough for good a few years before the Grand Concourse officially opened for business, and even had he stayed, he wouldn't have missed much.

Given the sense of anticipation that greeted the boulevard's arrival, the street's early years might have been expected to be eventful. And in fact, the hundreds of photographs taken by the city's Bureau of Highways to record construction of the road make the project look rather exciting and certainly ambitious, though living through seven years of digging and drilling must have been a nightmare for the occupants of the Victorian houses along the street's route, some of whom had to build plywood stairways just to venture beyond their front doors. In any case, whatever giddiness accompanied the building of the boulevard did not last. Photographs taken shortly after the official opening depict an expanse so desolate, so devoid of people, traffic, or other signs of life—not so much as a tree or even a bush to break the monotony—that the thoroughfare seemed already doomed. Judging by these images, it looked as if they had built a magnificent street and nobody bothered to show up.

Even during the next decade, signs of activity along the boulevard were minimal. With horse-drawn carriages increasingly a vestige of the past, the center portion was paved to allow for automobiles, marking the final step in the transformation of the Grand Concourse from turn-of-the-century greenway to urban thoroughfare. Yet cars were so rare in those years that watching a policeman with a whistle direct what little traffic existed counted as a major diversion, as did observing the antics of a locally celebrated cow tethered to a tree near Fordham Road, a reminder of the farms that still flourished in the area. "Youngsters [tried] to tease the cow, which paid no attention," an early resident named Esther Hoffman Beller told an oral historian at Lehman College many decades later. "And they mooed and mooed to the cow, and the cow went on chewing her cud." It wasn't exactly watching the grass grow, but almost.

To the few who ventured up to the West Bronx in those years, this lethargy might have seemed disappointing yet hardly surprising. During many of the boulevard's early years, the nation was either preparing for war, fighting a war, or struggling through the economic recession that followed the fighting. Some notable institutions had been erected in the Bronx, among them Stanford White's campus of the uptown branch of New York University overlooking the Harlem River. Yet even as construction barreled ahead in Manhattan and Brooklyn, development in much of the Bronx remained sluggish, in large part because the absence of subway service made it hard for people to move easily through the hilly terrain. Not until 1918, with the completion of the Jerome Avenue subway line, would this part of the West Bronx be made significantly more accessible.

A more intangible force also may have been at work. In the two decades between the road's conception and its realization, the world had changed in profound ways. The year the Grand Concourse was first envisioned, a primary audience was the young swells eager to race their horses along its cindery roadway. By the year of the boulevard's birth, technology had advanced so dramatically that the pioneer aviator Wilbur Wright took off from Governor's Island for what was the first flight over American waters. Although the street eventually caught up with its new role, the transformation did not take place overnight.

Gradually shifting mores affected even the way people referred to the boulevard. The word "Speedway" so beautifully inscribed atop Risse's original drawings had faded into memory, to be replaced with the name Grand Boulevard and Concourse, something of a mouthful even in that more formal era. In the 1920s, the Bronx County post of the American Legion and the Gold Star Mothers, whose sons had died in the First World War, pushed to have the street rechristened Memorial Parkway, honoring the dead of a conflict still fresh in memory. But their efforts were rebuffed, and Grand Boulevard and Concourse, shortened by future generations simply to the Grand Concourse, remained the official name.

By the early 1920s, as if coming out from under a spell, the boulevard began shaking itself awake. The city itself was exploding—a population that stood at about 1.5 million in 1890 leaped to more than 3.4 million by 1900, to 4.8 million in 1910 and 5.6 million by 1920—and the Bronx was following suit, with attendant changes in its most important roadway.

The Victorian houses along its route, with their gables, turrets, and wraparound porches, gradually gave way not to the elaborate villas that Risse had envisioned but to five-story and six-story apartment houses. These new buildings ranged from modest walkups for families of limited means to luxury structures such as the opulent Theodore Roosevelt, and they demonstrated what a powerful growth engine the boulevard was finally proving to be. By 1924, an official of the Bronx Board of Trade was describing the boulevard as "one of the foremost thoroughfares of the world, a thoroughfare which has added immeasurably to the prestige of the Bronx as a community and has created wealth for the owners of property along its length and adjacent territory."

Other changes took place to accommodate the arrival of a population for whom public safety was a more urgent issue. The decorative little electric lamps on the traffic islands, for example, gave way to more modern fixtures. The boulevard also grew a little longer. As the Grand Concourse became more popular and hence busier, the roadway was extended south to 138th Street to include a newly widened Mott Avenue, the street on which Louis Risse had spent so many productive years and Theodore Dreiser such unhappy ones, though perhaps in deference to its history, it was always described as a 4.5-mile strip.

=== • ===

If the name Grand Concourse had been uttered to the immigrant Jews from Russia and Eastern Europe disembarking on Ellis Island during the street's first decade of existence, they would have responded with a blank stare and a shrug. The name, like so many other points of reference in their newly adopted city, would have been meaningless. But by the early 1920s, as the broad boulevard in the West Bronx was fast becoming both destination and symbol, its reputation began to spread.

Social historians such as Irving Howe and Oscar Handlin have devoted countless eloquent words to describing the tormented lives of the Jewish immigrants whose first toeholds in America were the squalid and congested tenements on the Lower East Side, and to analyzing the forces that propelled them to neighborhoods farther afield after they got their bearings in the New World. It was, however, the actor Al Jolson, in the 1927 movie *The Jazz Singer*, who uttered one of the most passionate statements of the dreams and desires that drove these restless landsmen.

Despite its reputation as the first feature-length motion picture with synchronized dialogue, contemporary audiences typically find *The Jazz Singer* as primitive as the earliest silents. And for what was widely regarded as the world's first talkie, there's remarkably little actual talking. Yet among the few spoken sentences are those in which Jolson, portraying Jakie Rabinowitz, the cantor's son who longs for a career on the stage, describes to his mother the golden world that lies just beyond their horizon: "Mama, dahlin', if I'm a success in this show, well, we're going to move from here. Oh, yes, we're gonna move up to the Bronx. A lot of nice green grass up there, and a whole lot of people you know. There's the Ginsbergs, the Guttenbergs, the Goldbergs, and a whole lot of Bergs. I don't know 'em all." The Grand Concourse is never mentioned by name; the generation that achieved an address on the boulevard was the child's, not the parents'. Nevertheless, by the time the movie appeared, the beautiful street uptown had become an increasingly familiar destination for the city's immigrant Jews—and especially for their more affluent and Americanized offspring.

The journey that Jolson described had started as a dream in the minds of immigrants not long removed from the shtetls of Russia, Poland, and other parts of Eastern Europe, an impoverished flood tide fleeing repressive regimes with little sympathy for those of their faith. They began arriving in New York around 1880 and continued to do so for nearly three decades. Nor was the Bronx their only destination after getting their bearings on the Lower East Side—the East Side, as they called the neighborhood—or even their first. As early as the 1890s, immigrant Jews crossed the East River to establish communities in Williamsburg and later in the more distant Brooklyn neighborhoods of Brownsville and East New York, the pace of settlement quickened by the construction of the Williamsburg Bridge in 1903 and the Manhattan Bridge six years later. Others ventured north to Harlem.

The Bronx, however, beckoned most powerfully, first the walkups on Fox and Simpson streets in the decidedly working-class East Bronx, then, by the 1920s, as the population on the move became increasingly prosperous and adventurous, the sturdy apartment houses of the West Bronx, where janitors were called supers, where floors were covered not with linoleum worn thin from endless scrubbing but with glossy parquet—the "true middle-class country," as sociologist Samuel Lubell described that world. The name Grand Concourse grew more alluring by the day.

Families did not make the move in a single bounce. Rather, they leap-frogged from one apartment to slightly more impressive lodgings along what Lubell called "the old tenement trail," a route his own family traveled. Like a vast ragtail army, hundreds of thousands of immigrants and their offspring crisscrossed the city. But so many of them headed north, as if drawn by an invisible yet powerful magnet, that the Bronx during the 1920s emerged as New York's fastest-growing borough, the "wonder borough," as it was anointed in 1925, having absorbed half a million new residents within a decade. In February 1922, the *Bronx Home News*, the borough's influential daily newspaper, urged readers to "get a new resident for the Bronx" each day and adopted the slogan "one million population by 1925." With each passing week, the optimistic prophecies seemed closer to reality.

An expanding maze of new subway lines fueled this passage uptown. The mass transit that once followed the population now charted fresh patterns of growth and, especially in the West Bronx, served as the engine by which these newcomers colonized fresh terrain. The Jerome Avenue elevated—the IRT line completed in 1918 a few blocks west of the Grand Concourse—whisked riders to Manhattan in minutes; when the Concourse Plaza Hotel opened in 1923 on the boulevard at 161st Street, a promotional brochure touted its location as "fifteen minutes from Grand Central Station and thirty minutes from Wall Street." By the late 1920s, workmen were digging a trench under the Grand Concourse for the northern leg of the IND line, which opened in 1933 and provided fast and efficient access to the Garment District and other parts of the West Side.

=⇒ • ⇐=

Most of the newcomers during these years were Jewish, and for half a century the Grand Concourse and its surroundings stood as one of the city's most prominent Jewish neighborhoods. Still, Jews were not the area's sole occupants. Nearly a century earlier, Irish Catholics driven from their homeland by the potato famine had settled in the blocks around Fordham Road. Large numbers of them worked on the railroad and lived in the shadow of what came to be known as Fordham University; the institution, founded in 1841, occupied a set of imposing Collegiate Gothic buildings on a sprawling tract east of the Grand Concourse. Other Irish families lived in Highbridge, the hilly neighborhood west of the Grand Concourse, and at the northwesternmost tip of the Grand Concourse, a tiny community of Italian Catholics took root around Villa Avenue.

Like offspring of the university, a series of parochial schools sprouted on and near the boulevard to shape the minds of all these young Roman Catholics: Cardinal Hayes High School, Christ the King, Saint Philip Neri Church School, All Hallows High School, and the Academy of Mount Saint Ursula, later Ursuline Academy, notable for a Ring Day tradition in which seniors wearing navy blazers, blue oxfords, navy tams, and plaid skirts assembled at Poe Park and roller-skated shakily up the Grand Concourse, across Bedford Park Boulevard, and down the hill to their school, there to remove their torn stockings and bandage their bloody knees.

But Jews were the ones who made the Grand Concourse and its environs their own. Between 1920 and 1930, the number of Jews living on and near the boulevard increased by 450 percent; by 1930, 61 to 82 percent of residents in the Grand Concourse neighborhood were Jewish, making the area one of the city's five most heavily Jewish communities. Increasingly, the Grand Concourse neighborhood stood as a Jewish island in a Jewish borough; between 1920 and 1930 the Jewish population in the Bronx doubled, and by the decade's end, the borough was home to 585,000 Jews, the highest concentration in the city. For all these families, the Grand Concourse was the lodestone by which distances were measured. "Three short blocks from the Concourse," an advertisement in the *Bronx Home News* trumpeted of one apartment. "I'm right off the Concourse," bragged those who could. Even to live around the corner provoked murmurs of envy.

≡ • ≡

Of all the palatial structures that sprung up to house the affluent doctors, lawyers, and businessmen gravitating to the Grand Concourse during those years, few rivaled the white stucco behemoth known as the Theodore Roosevelt, which stood on the boulevard at 171st Street and was "what its builders claim is the largest apartment house in the world," as the *New York World* reported shortly before its opening in the fall of 1922. But the Theodore Roosevelt had an even more interesting claim to fame: its developer, Logan Billingsley.

New Yorkers were more familiar with Logan's younger brother Sherman, who as impresario of the Stork Club had created the fizziest symbol of the Jazz Age city, "the New Yorkiest spot in New York," as resident gossip columnist Walter Winchell described the nightclub on West Fifty-eighth Street. It was, however, the eldest Billingsley boy whose success in New York City capped the more unlikely journey.

Shortly after Logan Billingsley's birth in 1882 in Tennessee, his parents moved the family to a hardscrabble frontier town in the newly settled Oklahoma territory. A ramrod-straight six-footer with wavy dark hair and intense black eyes, Billingsley was only twenty-two when he was involved in his first homicide. A local girl claimed that he had gotten her pregnant, the girl's father came after him with knives in both hands, and Billingsley shot the man through the heart. In the years that followed, bootlegging kept Billingsley in constant trouble with the authorities, and by the early nineteen-teens, he had been in and out of jail so often—thirty days here, sixty days there, ninety days someplace else—it was impossible to keep track of either his transgressions or his whereabouts. According to his own estimate, his endless brushes with the law resulted in perhaps a hundred arrests for liquor violations.

As of 1915, Billingsley was in Seattle, running the city's largest bootlegging ring with two of his brothers, and it was Seattle that witnessed the most sensational episode in his criminal career. When two police agents raided one of the family's liquor warehouses, a watchman returned fire, and in the bloodbath that followed, all three men ended up dead. Although Billingsley dodged prosecution, by 1918 he was behind bars in a federal penitentiary. Escaping, he was arrested in Ohio; escaping again, he fled to Detroit and then to the Caribbean. The cat-and-mouse games with the cops ended only when Billingsley settled in New York and began remaking himself into the most solid of Bronx burghers.

Given the drama of his early years, one can't help wonder why a man so obviously enamored of risk and adventure chose to travel such a safe and respectable path. One afternoon a few years ago, Billingsley's son Robert, custodian of his father's sometimes dubious legacy, pondered that question in an empty conference room in the office of his Manhattan real estate firm. Surrounded by hundreds of newspaper clippings, letters, and other documents that traced the jagged arc of his father's life, Robert Billingsley was silent for a moment. Finally he ventured a response: "I think my father realized that in New York he could shed his old identity. The Oklahoma years had been an embarrassment. This was a journey to respectability."

LOGAN BILLINGSLEY
President
Bronx Chamber of Commerce

Logan Billingsley, the onetime bootlegger from Oklahoma who went on to serve as the developer of the Theodore Roosevelt on the Grand Concourse at 171st Street, "what its builders claim is the largest apartment house in the world," as the *New York World* reported in 1922, shortly before its opening. (Bronx Chamber of Commerce)

The Theodore Roosevelt's fourteen six-story buildings occupied an entire city block and featured Iberian touches, among them mansard roofs of Spanish tile topped with little flags. (Museum of the City of New York)

Such transformations were hardly unknown in early-twentieth-century America; Jay Gatsby, by no means the only scoundrel to end up in a Gold Coast mansion, embodied a familiar American type. Or perhaps the answer was even simpler. "Maybe," Robert Billingsley said of his father, "he just wanted to stay out of jail."

—— • ——

Whatever drove Logan Billingsley to abandon the bootlegging game, the Theodore Roosevelt was a dazzling piece of work that established him as a man to be reckoned with. With its Iberian touches, among them mansard roofs of Spanish tile topped with little flags that flapped cheerfully in the wind, the Theodore Roosevelt was not alone in bringing European dash to the Grand Concourse during those years. But in addition to being sort of Spanish, the Theodore Roosevelt also had the distinction of being immense. Its fourteen six-story buildings occupied an entire city block and contained three hundred suites. Photographs taken during construction give the impression that workmen are ripping up the entire West Bronx to make way for the new development.

The hyperbole that greeted the opening, although typical of the era, was as outsized as the building itself. "A remarkable advance toward perfection marks the erection of New York's most luxurious Apartment Hotel, the Theodore Roosevelt," a commemorative postcard gushed. "It is a majestic tribute to the art and ingenuity of man; a splendid accomplishment of modern architectural science. . . . All the improvements, all the accommodations which this modern age have produced have been incorporated in this classic structure."

So it seemed. Meals prepared in the dining hall or the on-site restaurant could be delivered via dumbwaiter. Apartments included maids' rooms, and valet service was provided. Armed guards patrolled the premises, even though some residents must have wondered what crime they were guarding against in this exceptionally tranquil neighborhood. Every day, a quartet of porters polished the brass fixtures in the lobby and scrubbed the marble floors. A nice touch was the absence of hallways, an architectural detail intended to increase the amount of privacy afforded individual families. In the interior courtyard, an elaborate Italian garden overflowed with flowers and shrubs and was dominated by a fountain and a fifteen-foot statue honoring the project's namesake, flanked by a sleeping lion and a sphinx. The statue proved such a hit that Billingsley commissioned the Italian sculptor who created it to produce another work for the courtyard, this time using as his model the developer's two-year-old son.

Billingsley's desire to honor his own at this particular juncture in his career was understandable because by the time the Theodore Roosevelt opened, its developer was emerging as a major player in borough affairs. Six years later he took over the faltering Bronx Chamber of Commerce and promptly transformed the organization into the city's most active business group. He was a fixture at seemingly every important local public function—"a builder who builds well," one admiring columnist wrote. Nevertheless, despite the glow of civic virtue that suffused Billingsley's later years, not all traces of his bad-boy past had disappeared. By the summer of 1934, just months after the repeal of Prohibition, newspapers were reporting that the ex-bootlegger was working for the West Indian Rum Syndicate, urging that rum tariffs be reduced and offering his own recipes for the beverage, among them "rum with ginger wine—warming, cures indigestion."

≈ • ≈

Logan Billingsley, the man who built the biggest apartment complex on the Grand Concourse, and Andrew J. Thomas, the one credited with designing its most innovative housing, had two things in common. Both started out poor, and both left their most memorable legacy atop that great ridge in the West Bronx. In virtually every other respect, Billingsley and Thomas could not have been less alike. Unlike the roughneck from the West who came East to shed a disreputable past and make money on the right side of the law, Thomas was an idealistic child of the New York City slums who was moved by the poverty he witnessed in his youth to create some of the city's most revolutionary housing projects. The two men differed in another important respect. Billingsley's constituency was the rich. Thomas cared intensely about the city's poor and its swelling middle class.

Thomas's name crops up regularly in treatises about housing innovations in New York City, and rightly so, because if he did not actually invent the term *garden apartment,* he unquestionably deserves credit for popularizing this innovative form of housing. Yet accounts of his achievements often fail to convey what a remarkable individual Thomas was, and how creatively he addressed the needs of those New Yorkers unlucky enough to pass their days without maids, valets, or hot meals delivered by dumbwaiter to their doorstep. And though he was famously prickly as a colleague, his contributions to housing in the city are immense.

Thomas's early years were bleak. Born in 1875 and raised on lower Broadway, he was orphaned at age thirteen. Two years later he went to work collecting rents for a real estate speculator, a job that afforded him a firsthand look inside the city's filthiest and most overcrowded apartments. Not long afterward, when employed as a timekeeper for a building contractor, he spent nights in makeshift offices working out construction plans. These experiences shaped him in profound and lasting ways. "I had never taken a drawing lesson nor had any technical schooling in architecture," Thomas was once quoted as saying, "but I knew what people needed to make them comfortable."

The passion he brought to his work seemingly knew no bounds. Louis Pink, one of the city's early advocates of public housing, said of Thomas, "Housing is his religion. 'What better religion could there be than housing?' he often exclaims." Early in his career, Thomas announced to the world at large, "I'll abolish every slum in New York if I can gain the attention and help of charitable organizations, the state and society. . . . I'll raze nine or ten blocks at a time until the entire city is rebuilt." When he died in 1965 at the age of ninety, his obituary in the *New York Times* noted that "his grayish blue eyes would light up with enthusiasm or blaze with impatience at anything that delayed his plans for changing the face of the city."

Yet this Jacob Riis of the landscaped courtyard was far more than a dreamy-eyed do-gooder. Perhaps more acutely than any of his contemporaries, he understood that while the garden apartment offered considerable social and aesthetic benefits—building less densely on a plot of land afforded residents not only more light and space but also a greater sense of "home"—a primary attraction of this concept was its profitability. As Pink summed up Thomas's achievement, "He learned that beauty pays."

Thomas had his greatest impact in Jackson Heights, the newly emerging neighborhood in Queens where he developed a series of artfully designed "garden apartments for wage earners," as the *Times* put it. At least for some wage earners: the Queensboro Corporation, the major financial backer of Thomas's developments in the borough, touted one project as a "restricted garden residential section of New York City," an unsubtle reminder that units were off limits to Jews, Catholics, and blacks. But his namesake project, at 840 Grand Concourse, is generally considered his masterpiece, "his most congenial work," according to architect Robert Stern, a latter-day fan, and "probably Thomas's greatest single building project."

Thomas Garden occupies an entire block between 158th and 159th streets. In design, the circlet of five-story buff brick apartment houses grouped around a sunken interior courtyard looks almost spare. Yet hidden within their embrace and occupying about half the site sat a small jewel, a Japanese garden offering a touch of Zen tranquility in what was fast emerging as one of the city's most densely developed neighborhoods.

The garden's charms prompted considerably over-the-top prose, even among visitors who arrived half a century later. "A goldfish pond represent[ed] the quietude that was expected to reign in this haven from the tensions of the outside world," Brian Danforth, an urban planner at Hunter College and an unabashed admirer of Bronx architecture, wrote in the 1970s. "This protective quality was reinforced through landscaping that included large Japanese lanterns along the walkways that intersected at the pond where slightly arched bridges carried the stroller gracefully over the pool. It was the ideal environment that could calm the soul of even the busiest New Yorker."

Among those whose souls it calmed was a young Bronx boy named Gabe Pressman, whose grandparents were among Thomas Garden's earliest residents. Half a century later, Pressman—long since established as one of the city's best-known television reporters—could not recall the name of the complex, if he had, in fact, ever known it. Yet he vividly remembered the image of himself as a child, standing next to the pool and gazing admiringly at the fish that swam languidly in its waters. "It was tranquil," Pressman recalled of the bucolic little enclosure to which he used to escape during visits to his grandparents' apartment. "It was an oasis of beauty."

Although Thomas Garden lacked such amenities as elevators, it offered many other benefits that combined practicality with aesthetics and reflected a sensitivity to the needs of residents, particularly harried young mothers, that was unusual for the period. A dumbwaiter carried baskets of laundry up to the roof to eliminate the need for unsightly clotheslines. Fire escapes were tucked into the corners of buildings to make them less visible. In bad weather, restless children could let off steam in a specially designed indoor playroom.

Under the original plan, Thomas Garden was to be sponsored by a consortium of labor unions and operated as a co-operative. Neither came to pass. Even before the first of the 170 families arrived in 1928, union sponsorship faltered, and the complex was acquired by John D. Rockefeller Jr., who saw investment in middle-income housing as a way to address the problem of the slums.

Thomas Garden's days as a co-operative also ended abruptly. Union-sponsored co-operatives for middle-income families flourished elsewhere in the borough during that period, notably in the North Bronx. But largely because tenant ownership was a new and unfamiliar concept for the working class, too few potential residents of Thomas Garden were willing to plunk down the money required to buy their apartments—a thousand dollars up front for a four-room unit—and so the building opened as a rental property. Thanks to new tax abatements created for such projects, however, rents were kept low. And fittingly, given the restrictions governing some of Thomas's earlier developments elsewhere in the city, the year Thomas Garden opened, the complex acquired a sister project in Harlem, in the heart of what was then New York City's most congested neighborhood. The nonprofit co-operative was called Paul Laurence Dunbar Apartments, in honor of the celebrated African American poet. And unlike Thomas Garden, which was restricted to white families, Dunbar Apartments was reserved exclusively for blacks.

⇒ • ⇐

The Theodore Roosevelt and Thomas Garden were undisputed standouts. But in the land of the five-story and six-story apartment building, the eleven-story apartment house at Clifford Place south of 175th Street—thirteen stories, if you counted up from the back entrance down the hill on Walton Avenue—was king, especially since that complex had been named after a signer of the Declaration of Independence and the borough's most illustrious son, and came equipped with a blue-canopied entrance guarded by white-gloved doormen and a pink marble lobby outfitted with fountains and carved stone dragons, not to mention a tenant list heavy on doctors, in those years as prestigious a tenancy as one could hope for.

The exact address of the Lewis Morris Apartments was 1749 Grand Concourse, but people rarely needed to be so precise. "It was the only place I knew," recalled Max Wilson, an East Bronx schoolteacher who discovered the building in the 1950s when he and his new wife moved into her parents' apartment, "where you jumped into a cab in the Bronx and said, 'Take me to the Lewis Morris,' and everybody knew where it was." For nonresidents, its allure was even more palpable. "As far as I was concerned," recalled Robert Esnard, a former deputy mayor who grew up around the corner, "the people who lived in the Lewis Morris were the richest people in the world."

An early advertisement for the project, dated September 4, 1923, was sandwiched between ads for buildings on West End Avenue in Manhattan, and fittingly so, since the Lewis Morris, with 270 apartments, was the closest thing in the Bronx to the classic West End Avenue apartment house. The ad notes that this "High Class Modern Fireproof Apartment" offered Otis elevators and suites up to seven rooms, along with "large laundry and steam drying room in basement, U.S. Mail chute on each floor, telephone service in each apartment, day and night switchboard service." Sweeping views were a particular draw: the Lewis Morris so towered over other buildings along the boulevard, residents of many upper-story apartments could see the faint outline of the Manhattan skyline from their windows. Residents of the twelve-room penthouse must have felt as if they were gazing on the entire city.

Perhaps most of all, the building's prestige derived from its great concentration of medical professionals. So many of its doctor and dentist tenants maintained offices on the lower floors that it was the rare West Bronx child of a certain era who did not have a tooth yanked or a needle stuck into an arm in a chalk-white office smelling sweetly of ether at Number 1749; its reputation as a setting for medical dramas was such that the high-pitched screams of prospective young victims approaching with their mothers could be heard a block away, or so residents of the Lewis Morris claimed.

Joyce Sanders, the daughter of a prosperous otolaryngologist, was eight when her family moved into the building in 1935, and she was by no means alone in sensing that Lewis Morris residents breathed a rarefied air. "People were kind of intimidated by our building," Sanders admitted nearly three-quarters of a century later. "We noticed it when we brought home friends, which is why we stayed in our little group a lot."

If you were the doctor's wife or the doctor's daughter, and especially if you were the doctor himself, a certain decorum was expected. "My father never carried groceries on the street," Sanders recalled. "My mother never wore a housedress." Although not every Lewis Morris family had servants, her parents had a couple in help, a man who drove her mother on shopping excursions to Manhattan and acted as the family's butler while his wife cooked their meals. Decades later, when Sanders pictured her parents, she saw a man driving a black LaSalle and smoking a large cigar and a woman lying on a turquoise velvet sofa on Saturday afternoons, listening to Milton Cross host the broadcast of the Metropolitan Opera over radio station WQXR.

One Lewis Morris alumnus who denies experiencing any sense of class distinctions was another doctor's child, the Pulitzer Prize–winning journalist and author David Halberstam, whose family lived in the building during the late 1930s and early 1940s. "Good God, no!" Halberstam replied emphatically a few years before his death when asked if he remembered 1749 Grand Concourse as a particularly showy place. "It was supposed to be grand, but it didn't seem all that grand to me."

This, however, is the recollection of a born iconoclast whose family left the Bronx when he was only eight. Far more typical are the memories of Lewis Morris alumni such as Harrison J. Goldin, whose doctor father shared an office for a time with Halberstam's father. Goldin's family moved to the building in 1937, shortly after he was born, and although he left to go away to college, he returned to his parents' apartment a few years later to start laying the groundwork for a political career in the borough. "Lewis Morris was definitely prestigious—the scale, the tenancy, everything," said Goldin, who went on to become New York City comptroller and a state senator. "If you lived in the neighborhood, you knew there was something different and quite grand about Number 1749. You knew it was special. Everybody knew it."

≡ • ≡

The Theodore Roosevelt, Thomas Garden, and the Lewis Morris were hardly the only residential buildings to rise along the Grand Concourse before the Great Depression brought a temporary halt to development. During the decade that they arrived on the scene, the years in which construction along the boulevard finally took off with a vengeance, they received furious competition from apartment houses representing a veritable United Nations of styles: Tudor, Gothic, Romanesque, Renaissance, Baroque, Spanish. Developers praised the up-to-the-minute amenities of these structure, among them eat-in kitchens, dining alcoves, tiled bathrooms and kitchens, medicine chests. Their real attraction, however, involved something subtler and more ethereal.

These European-flavored buildings seemed to have arrived in the West Bronx by way of a fairy tale. Romantic, whimsical, and eclectic, they were wonderfully suggestive of faraway places. The apartment house at 2665 Grand Concourse, at Kingsbridge Road, for example, was inspired by the Palazzo Medici in Florence, its Italianate nature suggested by hints of upper-story balconies. Glancing architectural references to towers and battlements graced the façade of Number 2700, just to the north, along with a doorway that was a dead ringer for a miniature palace.

So vast and ornate were some of the lobbies, with their moody dark wood paneling and accents of red velvet, that they had the look of medieval or Renaissance dining halls, "like Queen Elizabeth would swing through the building any minute," as one resident recalled. With their carved stone ornaments and pointed brick arches, with chimneys, gables, crenellations, and turrets that seemed to have arrived straight from the English countryside, not to mention Tudor-style half-timbered bays, shields surrounded by garlands, scallop shells with echoes of the Baroque, and rooftop ornaments recalling Greek temples, here a touch of Venetian stonework, there a little Rococo flower tracery, and all these ingredients delightfully and sometimes recklessly tumbled together—these buildings were hard not to love.

It was as if the garden-variety Bronx apartment house had been splashed with the luster of Europe's great palaces and castles. The Gothic-style chests and carved baronial chairs in the lobbies seemed to have come from an illustrated history book and inspired all sorts of fantasies. After Yankees games, a West Bronx boy named Marshall Berman used to make up stories about the buildings as he walked back to his family's apartment, imagining the Renaissance bankers and Chinese mandarins who in his mind's eye might have inhabited such exotic places.

These storybook structures arrived at an ideal moment. Notwithstanding the giddiness of the Roaring Twenties, Americans were sometimes in a melancholy mood, wearied by war and economic disarray. Despite the occasional battlement, these apartment houses were designed to project an aura of tranquility and in subtle yet emphatic ways to transport their occupants to another time and place, a better time and place. Their calming colors and rich historical accents harked back to what was remembered as a less tumultuous moment in history, especially if nobody thought too hard about the quasi-military purpose of all those towers and turrets. Interior garden courtyards made people feel sheltered within a safe and protected community, as if the place where they lived were enfolding them in an embrace. At a moment when the outside world seemed an unsteady place, all this was very comforting, very soothing.

3 ≡

"I Was Living in 'a Modern Building'"

THE CREATION MYTH for the style known as Art Deco has a breath-taking simplicity. It decrees that Art Deco was born during a single glorious moment at a single glorious event—the Exposition Internationale des Arts Decoratifs et Industriels Moderne held in Paris in 1925. The story is compelling and, like so many creation myths, not particularly accurate. Art Deco was incubated in cultural developments that long preceded the extravaganza along the banks of the Seine, and the term by which the style would be known came into vogue well after the last exhibitor had packed up and left town. But the 1925 fair did serve one crucial role. It offered the world an unforgettable glimpse of this sleek, insistently modern approach to design.

The event was a sumptuous undertaking, one that blanketed seventy-two acres on both sides of the river, attracted more than sixteen million visitors, and even after expenses put fifteen million francs in the coffers of the French government. For six months, pavilions displaying the works of upward of fifteen thousand designers, manufacturers, and department stores transformed the city into a huge emporium whose promotional drumbeat was echoed even by the Eiffel Tower, dressed up for the occasion with the futuristic logo of the Citroen automaker.

This relentless array of items to wear and admire and hang on the walls of one's house traced its roots to an incandescent moment in the city's history, the years before the First World War, during which "the foundations for the esthetics of the modern world were laid," as David Garrard Lowe describes the period in his book *Art Deco New York*. It was, Lowe continues,

> the Paris of the Fauves, "the beasts," as they were dubbed, when painters such as Henri Matisse and André Derain shocked the fastidious by using vivid, pure colors—green, orange, black—colors that would be intrinsic to the Deco palette. It was the Paris where with his "Les

Demoiselles d'Avignon" of 1907, Pablo Picasso emblazoned Cubism on canvas, a Cubism which would profoundly affect the form of Deco architecture. It was the Paris to which in 1909 Serge Diaghilev brought his Ballet Russes, whose exotic Oriental-inspired sets and costumes by Bakst and Benois opened French designers to an eclectic spectrum of influences ranging from Indo-Chinese bibelots, African tribal gods, and Mayan pyramids to the ravishing treasures of King Tut's tomb, all of which would become a part of the Deco visual vocabulary.

And if the fair did not exactly witness the invention of what future generations called Art Deco, it hardly mattered because the event had so profound an impact. Three thousand miles away, on the other side of the Atlantic, the shock waves generated by the fair reverberated almost instantly—in fashion, in jewelry, in home furnishings, and most powerfully in architecture. In Manhattan, the brilliantly stylized fabrics and shimmering brooches that had so dazzled visitors to the exposition seemed writ large in the design of such structures as the Chrysler Building, the Empire State Building, and the Chanin Building, places where every day armies of people crisscrossed lobby floors set with gleaming terrazzo and rode up to the clouds—literally, as in the case of the Chrysler Building's Cloud Club—in elevators faced with inlaid wood veneer and edged with stainless steel. A thousand miles to the south, in the waterfront community of Miami Beach, the style expressed itself in hundreds upon hundreds of ice-cream-colored bungalows and small hotels and apartments buildings whose tropical and nautical accents—palm trees, flamingos, whimsical flora and fauna, and ocean-liner motifs such as deck railings and porthole-style windows—suggested sunny little structures that seemed to have washed up from the sea.

It was, however, in the West Bronx, a few miles north of Manhattan, in nearly three hundred apartment houses—some forty of them directly on the Grand Concourse—built almost entirely between 1935 and 1941, that Art Deco left what many people consider its densest and most monumental imprint.

⇒ • ⇐

Even during the heyday of these buildings, not everyone loved them. "They were ugly as hell," the illustrator Edward Sorel, who grew up east of the Grand Concourse, once said of them dismissively. "Just cheap Deco." His contemporary Jules Feiffer, raised in the East Bronx, shared Sorel's disdain. "No sense of style," Feiffer said of the Art Deco buildings of the Bronx. "I found nothing attractive about that neighborhood."

Sorel and Feiffer were hardly alone in their opinions. Although these apartment houses were much admired in their youth, the style itself fell out of favor, and only in recent decades did a movement develop to protect and pay homage to the nation's trove of Art Deco. In the Bronx, another factor helped dim the luster of these buildings. Starting in the 1960s, so many of them experienced hard times that ever fewer traces remained of why they once caused such a stir. Often, it was hard to understand what all the fuss was about, especially when these structures were compared to exuberant Beaux-Art piles such as the Ansonia on the Upper West Side or the resplendent Apthorp Apartments, its Renaissance Revival neighbor a few blocks to the north. Some latter-day visitors trekking uptown to check out the Art Deco legacy along the Grand Concourse found the buildings so anticlimactic that they wondered if they had come to the wrong neighborhood.

Had these urban explorers taken a closer look, they might have changed their minds and realized that in their sometimes modest, sometimes flashy way, these structures are enormously beguiling, studded with details and flourishes that almost tumble over themselves in an effort to please. Although Art Deco apartment houses did not start appearing in the Bronx until after the Jazz Age was winding down, they came to epitomize the era's style and effervescence. Like the bobbed head, the rouged knee, the wailing saxophone, and the hip flask of bootleg gin, the Art Deco buildings of the West Bronx stand as indelible icons of Jazz Age New York. If you didn't exactly hear a Cole Porter melody as you headed home at the end of the day, at the very least your front door might be opened by a doorman wearing a smart uniform topped by a dashing little cape, and that door might lead to a jewel box of a lobby whose mirrored walls reflected a radiant terrazzo ceiling and a shimmering terrazzo floor.

≡ • ≡

So much has been written over the years about the Art Deco buildings of the West Bronx that it takes a monumental effort to remember the excitement they generated when they were new. Yet pretend for a moment that you are a prosperous doctor or lawyer or businessman who lived along the boulevard, and imagine how it felt to set foot in one of these apartment houses for the first time.

Art Deco being a style of ornament more than form, the shape before your eyes may not immediately knock you out. For the most part, the buildings are just five or six stories tall, faced with unassuming buff brick. Except for notable stunners such as the Fish building, near 166th Street and so called in honor of the brilliantly colored mosaics of marine life around the entrance, the façades can seem distinctly unprepossessing. Look more closely.

The façades are remarkable in two respects. One involves the imaginative way the architects used such devices as setbacks, recessed entryways, and rounded corners to break up the typically flat exterior wall. The other has to do with how architects manipulated mosaics, glass, metal, brick, concrete, and expensive wood such as teak and mahogany to underscore these effects and at the same time to appeal powerfully to the senses.

Thanks to all the decorative details and the piling on of luxurious materials, the allure of the façades lay in small and often subtle touches: the staggered arrangement of floors, the colorful brick detail wrapped around recessed windows, the brass and steel accents framing the entrances, the ziggurats and railings along rooftops that suggested Aztec or Mayan temples. It was as if a painter had embellished his canvas by sticking on chips of gems and bits of bright metal and shiny paper, then crisscrossed his creation with grids and starbursts and zigzags, and finally rolled the whole assemblage up onto the street for all to admire, both those lucky enough to live within its confines and those who simply strolled by. The buildings were theatrical, sometimes overtly so; slabs of precast concrete near the entrances suggested proscenium curtains that almost literally ushered residents into the bright spaces within.

If the façades looked like scrims, to enter the lobby often felt like stepping onto the stage itself. Here, designers drew on a repertoire of decorative elements that included terrazzo floors and ceilings, embossed elevator doors, invitingly plump couches, and statuary so ornate that the pieces seem purloined from a museum. In some lobbies, mosaicists and muralists clearly had a field day, as they did in the lobby of 1166 Grand Concourse at 167th Street, where a mosaic depicts a peaceable kingdom of birds, sea horses, and antelopes. Yet despite the charm of the façades and the lobbies, it was the apartments that spoke most eloquently to the inhabitants of these buildings, and in describing the apartments decades later, former residents invariably linger obsessively on the sunken living rooms and windows that wrapped around the corners of buildings.

The sunken living room is no more than a gently lowered area, reached by two or three descending steps, often with a little wrought-iron railing along one side. Nonetheless, those seemingly artless details give the illusion

The fanciful mural in the lobby of 910 Grand Concourse, an apartment house designed by Israel Crausman that also featured excerpts from the poem "Roofs" by Joyce Kilmer, in whose honor the park across from the Bronx County Courthouse was named. (Carl Rosenstein)

of a special and even dramatic space. Combined with the generously proportioned foyers typical of these apartments, the sunken living room makes already spacious quarters feel even more spacious and produces a sense of flow—a touch of feng shui in an otherwise no-nonsense part of the city.

Corner windows, made possible in the largest number of apartments by recesses in the façade, represent another modest design element that provides enormous punch. By facing onto two separate vistas, generally east or west, to allow in the most sun, these windows double the amount of light and air admitted by traditional windows. In an era before air-conditioning, the breezes captured by such cross-ventilation feel heaven-sent, and sunshine spilling in from two directions makes graciously proportioned rooms feel even more welcoming.

Even a brief visit to one of these apartments lingered in the mind, as it did for Elba Cabrera, who was born in Puerto Rico and raised in the Bronx and whose memory of a visit to the Grand Concourse apartment of

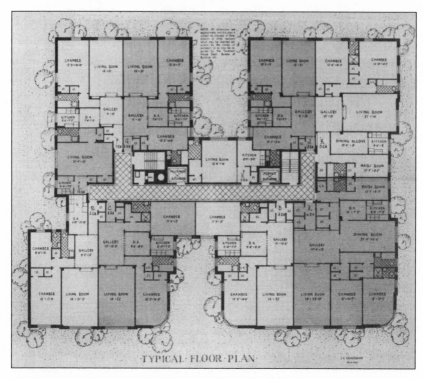

The floor plan of 910 Grand Concourse, showing many of the elements characteristic of the Art Deco buildings of the West Bronx, among them spacious foyers leading into sunken living rooms. (Bronx Art Deco Architecture/Hunter College)

a teacher named Mrs. Cohen stayed with her for decades. "I was so proud that she would let me visit her," Cabrera told an oral historian at Lehman College. "I remember the apartment, with the sunken living room and parquet floor. It was something I had never seen. I knew then that she had to be rich, because it was so different from what I was used to."

≡ • ≡

Paradoxically and tragically, the names of the men who created these buildings were largely forgotten within a generation, their legacy doomed by factors over which they had little control. They worked far from the heart of the city, and perhaps more crucially they lacked impressive pedigrees. Unlike the architects who came of age during the last half of the nineteenth century and were products of the same privileged ranks from

1001 JEROME AVENUE

A sketch of a façade of 1001 Jerome Avenue, a few blocks west of the Grand Concourse, featuring the wraparound windows that are a signature of local Art Deco apartment houses. (Bronx Art Deco Architecture/Hunter College)

which they drew their patrons, these men were largely Jewish immigrants or their offspring. For the most part, what education they received was acquired not at prestigious universities but at local art schools. Yet many of these men had deep roots in the borough where they produced so much of their work, and their energy, combined with their attachment to the Bronx, expressed itself in great bursts of productivity and creativity, notably in the area of housing.

Their numbers include Israel Crausman, a Russian who immigrated to the United States at age thirteen, studied engineering and architecture at Cooper Union, set up his own practice at age twenty-three, and maintained an office in the Bronx, where during his first seven years in practice he designed more than three hundred buildings. Charles Kreymborg, raised in the Bronx, learned drafting as an apprentice with a company that sold drawing materials and studied architecture as an apprentice in a firm where he never earned more than two dollars a week; he opened his own firm in 1921. William Hohauser designed two memorable buildings on Jerome Avenue, a few blocks west of the Grand Concourse, though his supremely gifted and prolific cousin, Henry, the premier Art Deco architect of Miami Beach, was the Hohauser who went on to enjoy the far bigger career.

One of the more gifted of this new generation of housing architects was Jacob Felson. Little is known of his background beyond the fact that he attended Pratt Institute and Columbia University and ended his career as a designer for the Tishman Construction Company. Nevertheless, his legacy survives up and down the Grand Concourse: standouts include Number 750, whose décor includes a lobby with a terrazzo floor and panels adorned with palms, and especially Number 1500, a particular favorite of *The AIA Guide to New York City*. "The parapet limestone has been carved into folds, like velvet," the guide's authors, Norval White and Elliot Willensky, write of the building, and "the glass block and stainless steel entry is wonderful." Wonderful is indeed the word for the effects produced by these blocks of translucent glass brick, which admit diffused light, and then, like a kaleidoscope tilted toward the sun, abstract the resulting images into a sparkling, ever-shifting collage.

Perhaps the sole architect who fit the mold only partially was the Hungarian émigré Emery Roth. One of the stars of his era, Roth left his mark primarily in Manhattan, where notable examples of his work include the twin-towered San Remo on Central Park West. He did, however, create a memorable apartment house in the West Bronx—at 888 Grand Concourse on the corner of 161st Street—a medley of curves, scallops, and concave spaces executed in polished black granite, bronze, stainless steel, marble mosaic, and gold stripes that was unique on the boulevard.

The most towering figure—"the genius of the Bronx," as one architecture historian called him—and an individual whose biography is far more complete than those of his less celebrated colleagues, thanks to the longevity of the firm he founded, the extent of its commissions, and the care with which his archive has been maintained, is Horace Ginsbern. Ginsbern

The façade of an apartment house on 201st Street off the Grand Concourse, another Crausman building. (Bronx Art Deco Architecture/Hunter College)

The lobby of 750 Grand Concourse, designed in 1937 by Jacob Felson, featuring a terrazzo floor and the use of luxurious materials that appealed powerfully to the senses. (Carl Rosenstein)

The interior of 888 Grand Concourse, opposite the Bronx County Courthouse. Its creator, Emery Roth, among the city's best-known architects, left his mark primarily in Manhattan, but he created one memorable building in the West Bronx. (*New York Times*)

The exterior of 888 Grand Concourse, with its distinctive circular floor and its concave entryway, was unique on the boulevard. (Carl Rosenstein)

and Associates, located at 205 East Forty-second Street, closed its doors in 1986, after nearly seventy years in business. But its Russian-born founder was one of the city's most prolific builders of apartment houses, and the firm's legacy in the West Bronx endures, both through its creations and the blueprints for those creations, which are preserved in the archives of Avery Architectural and Fine Arts Library at Columbia University, home to documents for 137 Ginsbern apartment houses in the Bronx, fourteen of them directly on the Grand Concourse.

These drawings were made for utilitarian purposes—to guide the contractors, plasterers, metalsmiths, and stonemasons who would bring form and life to two-dimensional pen-and-ink sketches. Yet to unroll the oversized, slightly stiff sheets of tissue and olive-gray linen (rejected drawings were cut up and turned into handkerchiefs) is an almost thrilling experience. The drawings are so studded with specific notations indicating exactly what materials were to be used and how they were to be employed that the structures depicted practically come alive under one's fingertips. Architectural blueprints are often works of art in their own right, but the Ginsbern drawings, in their delicacy and detail, are sometimes more beautiful than the real thing.

Even plans for the firm's less celebrated buildings, such as a six-story apartment house on the Grand Concourse at 167th Street, are alluring. The dimensions cited—a nineteen-foot-by-twenty-foot living room and twelve-foot-by-nineteen-foot bedrooms, along with a kitchen, a dining balcony, a gallery, and closets galore—indicate that the apartments were spacious and gracefully proportioned, and the aesthetic instructions make the public areas sound luscious. The blueprints specify that staircases incorporate hardwood handrails and marble treads, that the lobby feature "bronze saddles for elevator doors" along with accents of ornamental iron and stainless steel, and that the entrance hall be outfitted with a terrazzo floor, white Carrera glass, Belgian block marble, something called Duro-luxo wallpaper, a blue mirror with aluminum trim, and, over the mantle, a peach mirror with a wood frame. On the façade, where specifications call for slate and orange and white brick, the designs are drawn with such precision that you can see practically every brick.

These gorgeous documents seem to reflect the thinking of a cerebral aesthete, an individual who lived only to create beauty. Harry Ginsbern, as he was known, was not that man. Despite his formidable abilities, no one would have called him a dreamy artist; his grandson John Ginsbern, a Westchester County builder who worked for a time in the family firm,

2121 Grand Concourse, characterized by the zigzag façade and wraparound windows that were typical of Art Deco buildings in the West Bronx, was the work of Horace Ginsbern, "the genius of the Bronx," as one architectural historian called him. (Museum of the City of New York)

remembers his grandfather as a driven and demanding individual with a huge ego, a man as difficult as he was gifted. Ginsbern may not have been responsible for the façade of every building in the West Bronx that bears the name of his firm. Nevertheless, he was widely considered a dab hand at apartment layout, envisioning arrangements of interior spaces far more commodious than those found in the typical railroad flat. More important, he had the wisdom to choose as his design partner an immensely talented architect named Marvin Fine, whose skills complemented his brilliantly and who was almost as much of a mainstay of the firm as its founder.

When Ginsbern died in 1969, his brief obituary in the *New York Times* made no mention of his work in the Bronx. That is not surprising; by the time of his death, social and economic disarray was corroding much of the West Bronx, and having created anything in that part of the city was no great claim to fame. Nevertheless, it was Ginsbern, nearly as much as the boulevard's creator, Louis Risse, who imprinted on the Grand Concourse the face by which it would be remembered.

Details of Ginsbern's early life are sketchy and in some cases contradictory. According to his grandson, Horace Ginsbern was born Horace Ginz-berg in 1900 in a town near Minsk and was seven years old when he came to the United States. He studied at Cooper Union and Columbia University, and by the age of nineteen, he was married and the father of a son. Around that period, he opened his own firm, and at some point during the 1930s, he changed his name to the more American-sounding Ginsbern, though not in time to prevent him from being listed as Horace Ginzberg or Ginsberg in several reference works and even on the façades of some of his buildings.

A short man—five foot six would have been a generous estimate— Ginsbern had a round face, dark eyes, and a small matching mustache whose resemblance to Hitler's did not go unremarked upon. It was almost too tempting to speculate that his diminutive stature contributed to his feverish need to prove his worth. He worked hard, talked shop endlessly, and lived as hard as he worked, consuming great quantities of Scotch, often from the bar in his office, along with cigarettes and pastrami sandwiches. "He had a prodigious appetite for everything that was bad for him," recalled his grandson, who inherited the collection of silver dollars that his grandfather, an inveterate gambler, had won in Las Vegas. "And he had a very powerful sense of his own vision—in that sense he was the classic architect." Few who knew Horace Ginsbern would have been surprised that he died, of heart-related problems, at the relatively young age of sixty-nine.

Yet that outsize personality had its winning side. Although his colleagues' memories of his high-pitched voice bouncing off the walls of his office were all too vivid, when that voice was not striking terror in the hearts of his co-workers, those dark eyes twinkled. The energy and ambition that made him a millionaire several times over manifested themselves early on. And had Ginsbern been a gentler soul, he might have left a far less impressive legacy.

Along with the firm's work in the Bronx, it made a name for itself through an eclectic array of commissions in Manhattan. One of Ginsbern's earliest buildings, dating from 1921, was a storefront for a corset shop on Fifth Avenue. An apartment house of rosy tan brick on Amsterdam Avenue was hailed for helping provide "a new kind of city view . . . a little piece of new New York." And an office building on First Avenue at Sixty-first Street that was banded with contrasting strips of brick was featured decades later in an exhibition titled "Landmarks of the Future." For years Ginsbern was the architect for the Chock Full o'Nuts chain, and his string of Hanscom Bake Shops, with their apple-green enamel panels, curved corners, porthole doors, and banded windows, were deemed "admirable" by no less a critic than Lewis Mumford, who also praised a Ginsbern apartment house at Second Avenue and Sixty-first Street as "an impeccable example of good manners."

Despite a lack of formal training, Ginsbern served as the lead architect on Harlem River Houses, built in 1937 just north of Andrew Thomas's Dunbar Apartments. Like Thomas's project, the complex of low-rise redbrick buildings was a groundbreaking development, the first public-housing project financed, constructed, and owned by the federal government. Built specifically to accommodate black families, Harlem River Houses was praised by the *WPA Guide to New York City* as a complex "whose very success made the plight of the less fortunate residents of Harlem seem by contrast more bitter than ever." Nevertheless, Ginsbern and company left their sharpest and most memorable imprint on and near the Grand Concourse, beginning in 1931 with the opening of Park Plaza Apartments—"a pioneering work which helped change the face of the borough," the New York City Landmarks Preservation Commission concluded half a century later.

⇛ • ⇚

Park Plaza Apartments, four blocks west of the Grand Concourse, at 1005 Jerome Avenue near 164th Street, may not have been the first Art Deco apartment house in the Bronx. It was, nevertheless, unanimously regarded as the most influential, and the exuberant façade offered the first major example of the creative abilities of Ginsbern's partner, Marvin Fine, who had joined the firm a few years earlier.

The two must have made an unlikely pair. Fine, a gentle and affable man with reddish-blond hair, was in appearance and personality the exact opposite of his longtime boss, and nearly as interesting a figure. Unlike Ginsbern, Fine was a native New Yorker, born in Harlem in 1904, and also unlike Ginsbern, he received a first-class professional education, studying Beaux Arts architecture at the University of Pennsylvania. After graduation, he went to work for the celebrated Cass Gilbert, designing gargoyles for Gilbert's New York Life Insurance Building. Eighteen months later, by then sick to death of gargoyles, Fine joined the Ginsbern firm, where he rose to senior partner and stayed nearly his entire career. Notwithstanding their differences in temperament, or more likely because of them, the partnership endured for decades and allowed the firm to produce probably the finest and certainly the most numerous examples of the Art Deco that defined the look and mood of the Grand Concourse for decades to come.

Despite the change in Fine's professional address, he remained enamored of the Beaux Arts tradition, and his early drawings for the façade of Park Plaza Apartments show a riot of urns, swags, and other classical motifs. At the same time, he was much taken with the skyscrapers springing up downtown, especially Raymond Hood's American Radiator Building, near the New York Public Library, and William Van Alen's Chrysler Building, which was rising only a block from his office. These romantic towers were clearly in Fine's mind as he contemplated the apartment house his firm was building a few blocks west of the Grand Concourse.

When construction started in January 1929, the firm issued a press release touting the unusual size and scale of the project, a ten-story building that would house a total of 250 families in apartments of up to five rooms. Mezzanine arcades would overlook entrance halls twenty feet high. As for décor, the announcement noted that "the use of polychrome terra cotta [would blend] harmoniously with the light brick used and a very pleasing effect will be created."

On June 25 of that year, almost the very moment Park Plaza's doors were to open, tragedy struck: the building was "swept away by a spectacular fire shortly before midnight." Fine had gone to a Broadway play that evening with his parents, who were celebrating their anniversary, and upon leaving the theater, he noticed a headline in an "extra" edition of a news-paper about a terrible fire raging in a Bronx apartment house. Recognizing his building, he raced uptown to inspect the damage, but so huge a crowd had gathered to watch the blaze that he could not even make his way to the site to witness the destruction of his first masterwork in the Bronx.

Rebuilt in smaller and more subdued form, Park Plaza Apartments opened two years into the Great Depression, hardly the ideal moment to bring new housing onto the market. Yet even in modified fashion, the complex was impressive, particularly the colorful terra-cotta images on the façade, a medley of garnet, burnt orange, silver, and grassy green depicting a fountain flanked by flamingos, the sun rising from behind the roofs of Bronx apartment houses, and an architect presenting a model of his new building to the Parthenon for approval, an unsubtle reminder that every-one, from Park Plaza's creators to its first residents, should feel proud to be associated with this monument to the future.

So began the flood of Ginsbern buildings in the West Bronx. The same year saw the arrival of Noonan Plaza, a three-hundred-apartment, Mayan-inflected complex at Nelson Avenue and 168th Street, half a dozen blocks west of the Grand Concourse. At Noonan Plaza, so elegant that its doormen were attired in uniforms adorned with small capes, Ginsbern and company literally pulled out all the stops. Residents lucky enough to occupy apartments that overlooked the fifteen-thousand-square-foot in-terior garden were treated to a spectacular tableau: a waterfall splashing into a pool that was home to swans, goldfish, and water lilies and, in a tip of the hat to Thomas Garden, crisscrossed by a series of Japanese-style bridges. Amenities inside included mothproof storage closets, Electrolux refrigerators, and an elevator roomy enough to accommodate wheelchairs for the elderly. Bathrooms featured built-in tubs, colored tile walls, and hampers—all luxuries in their day. The promotional brochure included an "Honor Roll of Craftsmen" that proudly listed the names of all the skilled workmen who had labored on the building. And the brochure seemed merely to be stating facts in proclaiming that Noonan Plaza was meant to be "one's permanent home . . . designed to remain ever free from mediocrity."

Half a century later, the firm's legacy in the West Bronx was still draw-ing raves. The *AIA Guide* singles out Town Towers at 2830 Grand Con-course—"Its brick piers and crenellated parapet shimmer hello as thou-sands drive by it. Take a peek at the spectacular lobby"—and especially 2121 Grand Concourse—"the most stylish on the Concourse . . . Art

Park Plaza Apartments at 1005 Jerome Avenue, a 1931 work by Horace Ginsbern and his design partner, Martin Fine, that is generally considered the first Art Deco apartment house in the Bronx. Half a century after its completion, Park Plaza was hailed by the New York City Landmarks Preservation Commission as "a pioneering work which helped change the face of the borough." (Carl Rosenstein)

The façade of Park Plaza Apartments is rich with images of polychrome terra-cotta that include the frozen fountain, a popular Art Deco motif, flanked by flamingos and the sun rising from behind Bronx apartment houses. (Carl Rosenstein)

At the entrance to 1035 Grand Concourse, another Ginsbern building, bold stripes and porthole accents offer a touch of Miami-style Art Deco. (Bronx Art Deco Architecture/Hunter College)

Zigzags and setbacks on the façade of the Ginsbern building on the Grand Concourse at 184th Street; the architect, who signed his name as Horace Ginsberg, eventually changed it to the more American-sounding Ginsbern. (Carl Rosenstein)

Deco at its best." But the Fish building—no one ever called the apartment house at 1150 Grand Concourse anything else—was hands down everybody's favorite. Bronxites routinely used words such as "astonishing" and "astounding" to describe Ginsbern's splashiest creation, and they insisted, probably correctly, that nothing similar existed along the boulevard or anywhere in the city.

The small gem of a lobby, a circular space enclosed by wood-paneled walls, featured a marble fireplace, recessed lighting fixtures, elevator doors accented with red and metal inlay, and a starburst terrazzo floor of reds, golds, and greens. On facing murals, weirdly pale and elongated maidens struck languorous poses against a background of swans, frogs, and castles.

As if setting the scene for the mosaics flanking the main entrance, the façade sparkled with aquatic touches. Overlapping three-quarter circles of stainless steel on the front doors suggested schools of fish, the half circles on the entablature resembled stylized waves, and a person peering long enough at the green glass slabs inset in the entry area had the momentary and not unpleasant sensation of being under water. The main event, however, was the mosaics themselves, a fantasia of marine life executed in chips of stone in every color of the rainbow. Wide-eyed jellyfish and other aquatic creatures swam lazily through a tropical sea of bubbles, wavy lines, and tendrils of exotic plants. The charming, slightly goofy images were certainly the last thing anyone expected on this solid, upper-middle-class street. And everyone adored them.

≡ • ≡

For the great majority of the families who lived in these forward-looking apartment houses, the main emotion they evoked was uncomplicated pride; how could a person not revel in lodgings that both reflected and magnified one's own sense of worth and achievement? Yet among some residents, these structures inspired more complex feelings, a profound sense of partaking of a singular moment in history.

The political scientist Marshall Berman, who grew up in the West Bronx in the 1940s and 1950s, felt these emotions intensely. "For most of my life," Berman writes in *All That Is Solid Melts into Air*, his 1982 meditation on Modernism and its power, "since I learned that I was living in 'a modern building' and growing up as part of 'a modern family,' in the Bronx of thirty years ago, I have been fascinated with the meaning of modernity."

For Berman, the Art Deco buildings of the West Bronx represented the same thrust toward an urban future expressed in the great glittering boulevards with which Baron Haussmann, under the patronage of Napoleon III, had replaced the narrow streets that characterized medieval Paris. "I can remember standing above the construction site for the Cross-Bronx Expressway," writes Berman, who lived on College Avenue, a few blocks east of the boulevard.

> The Grand Concourse, from whose heights I watched and thought, was our borough's closest thing to a Parisian boulevard. Among its most striking features were rows of large, splendid 1930s apartment houses: simple and clear in their architectural forms, whether geometrically sharp or biomorphically curved; brightly colored in contrasting brick, offset with chrome, beautifully interplayed with large areas of glass; open to light and air as if to proclaim a good life that was open

The small gem of a lobby in the so-called Fish building, a Ginsbern creation at 1150 Grand Concourse. The floor was a terrazzo starburst of golds, greens, and reds, and the elevator doors were accented with red and metal inlay. (Carl Rosenstein)

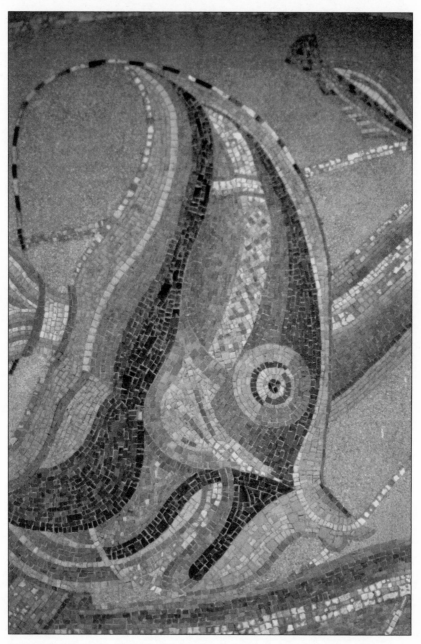

The multicolored mosaics on the façade of the Fish building, depicting giant, wide-eyed fish and surreal aquatic plants, were unique on the Grand Concourse and probably in the city. (*New York Times*)

not just to the elite residents but to us all. The style of these buildings, known as Art Deco today, was called "modern" in their prime.

Berman was by no means blind to Modernism's dark side; he was witnessing one of its fiercest manifestations that very day as he contemplated preparations for the expressway soon to plow through the valley atop which Louis Risse's "Great Wall of China" had been erected. But neither was he immune to Modernism's allure. "For my parents," Berman continues, "who described our family proudly as a 'modern' family, the Concourse buildings represented a pinnacle of modernity. We couldn't afford to live in them—though we did live in a small, modest, but still proudly 'modern' building down the hill—but they could be admired for free, like the rows of glamorous ocean liners in port downtown."

=== • ===

The Art Deco apartment houses along the Grand Concourse were the most dramatic features of the West Bronx landscape. They did not, however, exist in isolation. Even on the boulevard proper, they did not form an unbroken row but rather sprouted in clumps, as if nourished by patches of especially fertile soil. Interspersed with them, and lining the side streets, were great stretches of more modest five-story and six-story apartment houses. These structures, occupied by legions of small shopkeepers, civil servants, and garment-center cutters and operators, blended with their illustrious fellows to form a thickly woven tapestry.

Although this dense streetscape emerged as a hallmark of the West Bronx, similar neighborhoods existed elsewhere in the city, notably along Eastern Parkway and Ocean Parkway in Brooklyn and Queens Boulevard in Queens. For countless immigrant Jews and their offspring, such districts performed two important functions. They allowed escape from the squalor and congestion of the ghetto, providing a critical next step on the transformative immigrant journey, while at the same time replicating aspects of the urban experience that, often unconsciously, these newcomers had come to cherish.

Not for these immigrants or even their children was the dream house a cottage atop a private patch of grass; the prospect of such a dwelling was almost terrifying, no matter how bleak the alternative. Michael Gold, author of the 1930 novel *Jews without Money*, a pitiless depiction of the grimness of tenement life, understood this fear profoundly. Among the book's most wrenching scenes is the one in which the protagonist's father accompanies his boss to the emerging Brooklyn neighborhood of Borough Park to inspect a newly built single-family house:

> My mother would not go in. She remained on the porch like a beggar. With troubled eyes she stared around her at the suburb, at the lots covered with slush and weeds, at the eight banal houses.
>
> I followed my father into the new house, with its bleak smell of varnish and wood shavings. I heard my father's Boss chant:
>
> "Hardwood floors, Herman! A first class kitchen range free! Electric lights! A modern water closet! Oy, what a water closet! Only America has such water closets! . . .
>
> On our way home, my father asked my mother: "Well, what do you think of it now, Katie?"
>
> "I don't like it," said my mother.
>
> "And why not," my father said indignantly. "Are you so much in love with that sewer of an East Side?"
>
> "No," said my mother. "But I will be lonesome here. I am used only to plain people; I will miss the neighbors on Chrystie Street."
>
> "But there will be neighbors here," said my father.
>
> "Herman, don't make me do it," my mother pleaded. "I can't do it, Herman. My heart is heavy thinking about it."

Jews such as these, accustomed to messy yet intense confinement in the heart of a vast metropolis, regarded a freestanding private house not as blessing but curse. In their eyes, there was something infinitely worse than a life of poverty and filth—a single-family dwelling with modern plumbing and an up-to-date stove.

Nor, despite the shriek of pain that is Gold's novel, did closeness necessarily represent unrelieved misery. "The crowded tenement neighborhoods spontaneously generated associations," Oscar Handlin points out in *The Uprooted*, his classic study of the American immigrant experience. "Here people could not help meeting one another, their lives were so much in the open, so much shared. The sounds of joy and sorrow traveling up the airshaft united all the residents of the house; the common situation that cut these men off from the rest of the city itself united them. Within the ghetto could grow understanding, then sympathy, and in time co-operation."

Out of this desire to replicate the pulsing life lived on the Lower East Side was born the Jewish love affair with the apartment house and, in turn, with neighborhoods such as the West Bronx where such housing prolif-erated. This world also offered its own grandeur. The apartment house, suggests Columbia University professor of architecture Richard Plunz, be-came a "people's palazzo, involving a monumentality not possible through individualism. The cottage yard with grass and flowers became a courtyard instead, elaborately landscaped and collectively maintained for the com-mon good."

Social scientists have repeatedly been struck by the Jewish affinity for the apartment house, a feeling that reached a crescendo at the exact mo-ment countless numbers of Americans yearned for its opposite. To be sure, not every suburban neighborhood welcomed Jews with a wide em-brace. Yet the Jewish love affair with the urban apartment building seemed powerful and deeply rooted.

"Wherever Jews have formed an important segment of the urban Amer-ican population, they have been the prime developers and prime residents of apartment houses, especially of the so-called 'elevator buildings,' those bastions of the urban middle class," Brandeis University sociologist Mar-shall Sklare wrote in *Commentary* in 1972. "Such construction reached its apogee along the great boulevards of Brooklyn, the Bronx, and, more re-cently, Queens—Eastern Parkway, Ocean Parkway, the Grand Concourse, and Queens Boulevard. These were essentially 'Jewish' avenues, built by Jewish developers for a Jewish clientele." Or as Plunz describes these pat-terns of growth, "The Grand Concourse and Pelham Parkway became the twentieth-century equivalents of Chrystie and Forsyth Streets; Third Avenue and East Tremont were the equivalents of Orchard Street."

Residents of these apartment buildings probably never expressed their preferences in quite such terms. Perhaps they never understood, except vaguely, why they felt so at home in their new surroundings. If asked to explain why modest five-story and six-story apartment houses so suited them, they might simply have stated the obvious, which is that the apart-ments offered considerably more in the way of light, space, quiet, and bathrooms than their previous accommodations had.

What could be more desirable, they might have added if pressed, than being able to rent a four-room apartment only blocks from the Grand Concourse for just a quarter of one's salary, not to mention a free month's rent upon moving in? Nevertheless, in some visceral way, these families understood the contours of the urban environment they craved, and when they came upon a setting that spoke to them, they recognized it immedi-ately and intuitively.

The fictional Molly Goldberg, who for decades and in various media represented to America the quintessential Bronx housewife, embodied these attitudes in vivid fashion. When she called out the window of her apartment on East Tremont Avenue to Mrs. Bloom, the woman living across the courtyard, she was miles and worlds away from what her audience would have imagined as the pushcart-lined streets of her younger days. But to be within yoo-hooing distance of one's neighbors stood as a reminder that at least one familiar ritual of an older way of life had endured and would continue to provide comfort.

PART II

The Golden Ghetto

4 ⇐

"Something That Everybody Had in Awe"

IN THE MID-1950S, the artist Franz Kline used the phrase "Miss Grand Concourse" to express his disdain for Ruth Kligman, the sexpot from New Jersey who was Jackson Pollock's lover, the sole survivor of the car crash that hurled Pollock's body into a tree and, with what everyone who hung out at the Cedar Tavern in Greenwich Village regarded as unseemly haste, the lover of Pollock's great friend and fellow Abstract Expressionist Willem de Kooning.

Kline was not the only person to speak dismissively of Kligman. The poet Frank O'Hara famously dubbed her the "death car girl," and de Kooning's wife, Elaine, called her Pink Mink. Yet even at a moment when the boulevard's luster was fading, Kline was atypical in twisting its name into an insult. Almost universally, and for decades, men and women who came of age on or near the Grand Concourse spoke of the street with an affection that at times bordered on besottedness. Almost automatically, they called it the Champs Elysees of the Bronx, sometimes the Park Avenue of the Bronx, and although a half smile may have flickered on their lips in acknowledgment of the incongruity of the image, the warmth in their voices was palpable.

In part, these men and women were responding to the street's undeniable beauty. The eleven-lane expanse was so broad, so perfectly situated on its long, narrow bluff. You could see so much sky. To suddenly find yourself in this huge, bright space in the heart of a tightly packed metropolis was a liberating experience, especially for those who came upon the boulevard unexpectedly and unawares, as was often the case. "I was stunned," Bronx historian Evelyn Gonzalez recalled of her first sight of the Grand Concourse. "I didn't have a clue that such a place existed." As a child coming of age in Spanish Harlem in the 1940s and 1950s, Gonzalez rarely ventured beyond the projects in which she was raised, and to her young eyes, the Grand Concourse emerged like a glittering Oz from a sea of drab five-story and six-story walkups.

The Grand Concourse looking south from Mosholu Parkway in 1932, a few years before Art Deco apartment houses began rising along the street's flanks. (Bronx County Historical Society)

The critic Irving Howe, who grew up in the East Bronx in the 1930s, wrote dreamily of the summer nights that he and friends briefly escaped to the boulevard from their families' drab and crowded apartments, the evenings they would "parade across the middle bulge of the Bronx from the tenements on Wilkins Avenue in the East to the forbidding apartment houses of the Grand Concourse on the West. I remember those night walks as carefree and relaxed, away from the pressures of family and poli-tics. . . . The streets would be empty, the summer nights cool, a kind of expansiveness would come over us. . . . We would glide away in our Mel-villian freedom."

Even those who discovered the street in its later, less impressive years were struck by its singular qualities. "There was such a sense of expansive-ness," recalled Mary Childers, who was the daughter of a welfare family whose home was a warren of dark, roach-infested basement rooms on Webster Avenue off to the east and who discovered the boulevard in the 1960s when she babysat for a family that lived on the Grand Concourse.

The Grand Concourse looking north in 1966. Even in winter, and even at a moment when its glory was fading, the street exuded a sense of majesty. (*New York Times*)

Childers was especially struck by how people dressed, not in the jeans and T-shirts that prevailed in her neighborhood but in freshly ironed dress shirts that tucked into waistbands and well-tailored slacks that required dry cleaning. "Nothing too low-cut or too tight," she recalled, "unlike what you saw on poor women, who couldn't afford to buy new clothes when they gained weight." Wandering about in a daze, she admired the necklace of parks and buildings whose fanciful decorations looked to her like frosting on a cake. "I felt like Nancy Drew, like a character in a story where someone clunks you over the head, and you're in a different world," she said. "I couldn't believe there was such a beautiful place so close to where I lived. Why didn't somebody tell me?"

Paradoxically, or perhaps not, some of the people most taken with the street were those, like Evelyn Gonzalez and Mary Childers, who experienced the boulevard from a distance, both literal and spiritual. This phenomenon helps explain why Bronx residents of even modest means felt such a keen sense of ownership, despite the street's powerful association with the well-to-do. "The Grand Concourse . . . was our boulevard," a Bronx secretary named Frances Rosenblatt told an oral historian at Lehman College. "It was something that everybody had in awe, really amazing. . . . And that was where all the doctors were, and the dentists, and the professional offices, [the people who] were more financially able."

A person could live not far away and never know anyone with an address on the boulevard proper. Paul Pevonak, son of a railway clerk, used to play softball on summer nights under the illuminated Chevy sign near Saint Ann's Episcopal Church in Mott Haven, near his home. Many years later, Pevonak confessed wistfully that his family never knew any of the doctors or lawyers or dress manufacturers who lived on the Grand Concourse, and indeed, how would they have?

Like a glittering medieval village off to the distance in the Italian countryside, the farther from the street one lived, the more sharply its nature came into focus. If your home was a walkup on the Grand Concourse, the elevator buildings across the way seemed the ultimate in apartment living. If you lived off the Grand Concourse, everything on the boulevard seemed touched with magic. And if you lived in the East Bronx, anything in the West Bronx, anything at all, seemed desirable.

⇒ • ⇐

Was the Grand Concourse inspired by the Champs Elysees? No one really knows. Louis Risse does not mention the Champs Elysees, even in passing, in his detailed description of the thoroughfare he envisioned in the West Bronx. Beyond the fact that Risse was a Frenchman who knew the Champs Elysees well from his youth, and beyond the superficial resemblance between the two streets, with their sweeps of roadway and sidewalk demarcated by seemingly endless rows of trees, no evidence exists that the great Parisian boulevard was in Risse's mind as he set about creating his own masterwork. Yet, whatever the engineer's intentions, the two streets share a great deal beyond mere beauty, namely, a more ineffable quality that has to do with their singularly urban environs. Like precious gems enclosed within fine settings, both streets were enhanced by the grand buildings that flank them.

In this respect, the Grand Concourse has more in common with the Champs Elysees than with Eastern Parkway and Ocean Parkway, the two Olmsted and Vaux boulevards in Brooklyn to which it is often compared. As major destinations for the city's upwardly mobile Jews, all three streets provided the spine of some of New York City's most iconic Jewish neighborhoods. Yet at heart, the Grand Concourse and the Brooklyn boulevards are very different creatures, born of sharply different worlds.

Eastern and Ocean parkways were conceived and built in the late 1860s and early 1870s, at a moment when nothing moving faster than eight or ten miles an hour would travel along them. Both were the work of landscape architects who viewed their creations as linear parks, "sylvan roads" on which to enjoy "pleasure drives through fields and gardens"; only later did they evolve into residential avenues.

By contrast, though envisioned just a quarter of a century later, the Grand Concourse took shape in what was fast becoming a much more urban city, one in which automobiles became an increasingly commonplace sight. Henry Ford's Model T, which made its first jerky journey in 1896, started bouncing off the assembly line in 1908, just a year before the completion of the Grand Concourse. By the time of its opening, the boulevard had already been retrofitted to accommodate these new and speedier vehicles. And unlike Olmsted and Vaux, Risse did not view the city's future from the perspective of a landscape architect; despite his remarkable gifts, he approached all his projects, and certainly the great work of his life, as an engineer, not an aesthetician.

The Grand Concourse differed from the Brooklyn boulevards in another respect. Except near their starting points at Prospect Park, where apartment buildings sprung up, the Brooklyn parkways sliced through a fabric stitched largely of four-story brownstones and other single-family houses and low-rise buildings, all set well back from the street. By contrast, the Grand Concourse sat snugly within a tight frame of five-story and six-story apartment houses, many of which stood side by side and practically hugged the sidewalk. That they were of uniform scale, their façades set smack against the street, gave the boulevard the air of being a single continuous corridor.

These buildings were the ultimate product of their age. Although Risse had envisioned villas along the boulevard, history provided something that proved far more invigorating, block upon block of handsome apartment houses. The result was a singularly rich and intensely urban environment, reminiscent not just of the Champs Elysees but also of the great boulevards of sophisticated European capitals such as Berlin and Vienna, lively settings that defined and gave stature to the cities in which they were located.

Along the Grand Concourse, street and streetscape intertwined, braiding together in potent and seductive fashion the life of the boulevard and the lives conducted within the buildings along its edges. A child taking an early-morning bicycle ride down the boulevard could drink in the intoxicating aroma of freshly made rolls drifting from the bakeries on the side streets; the same child heading home at the end of the day could smell the chicken soup and brisket that other children's mothers were cooking for dinner, the aromas drifting out of the windows facing the avenue.

⇒ • ⇐

Perhaps most important, the street functioned as a highly public stage for events major and minor. Speaking of the Grand Concourse and sister streets such as the Olmsted boulevards in Brooklyn, former *New York Times* architecture critic Paul Goldberger described them as "among the grandest urban gestures in this country." "These places constitute a true public realm," Goldberger continued.

> In the great era of New York as a middle-class city, a time that ran from the late 19th century to around the end of World War II, the city of the middle class was not merely one of rows and rows of little houses and little yards, each set far apart from the other. Neither was it rows of brownstones squeezed tightly together. There were plenty of those, and there still are, but there were also open spaces, stately avenues and squares among the individual buildings, a commitment to a kind of shared grandeur. . . . And it was the commitment to a genuine and noble public realm that made New York's outer boroughs so different from other places.

The Grand Concourse shimmered with special vibrancy on the Jewish High Holy Days, especially on Yom Kippur, when the prohibition against driving reduced the din and bustle of traffic to barely a murmur. "In the mind's eye, it was always the holidays which made the Grand Concourse a special place," *New York Post* reporter Arthur Greenspan, who grew up in the West Bronx, wrote of these occasions. On three days in early autumn, the sidewalks were virtually obscured by families heading home from temple—women wearing glossy mink stoles and men in sleek black Homburgs carrying their prayer shawls in blue velvet containers embroidered with the gold Star of David. The tableaux on the street seemed to mirror the rituals that had been enacted in the synagogue just moments earlier, the choruses of *Gut Yontif* and *Shanah Tovah*—"Happy Holiday" and "Happy New Year"—a secular echo of the words that had been recited by the rabbi from the Torah as he stood at the bimah.

Marchers parading along the Grand Concourse in the 1920s. Thanks to the boulevard's width and the underpasses that kept away crosstown traffic, the street quickly emerged as the borough's main parade route. (Bronx County Historical Society)

Thanks to the boulevard's width and the underpasses that allowed crosstown traffic to flow without interrupting activities on the main road-way—fewer than half the number of underpasses originally envisioned but enough to keep east-west traffic to a minimum—the street also emerged as a magnet for official processions. By the early 1920s, the Grand Concourse had established itself as the borough's main parade route. These were patriotic years, shadowed by the memory of what many still called the Great War, and the annual procession celebrating Decoration Day, as the Memorial Day holiday was known in the early decades of the twentieth century, was one of the great Bronx institutions, a blur of crimson and gold uniforms, stiff hats adorned with feathers, artificial poppies stuck in lapels, and brass instruments that flooded the May air with stirring martial music. In 1921, the local American Legion post planted 952 young maples along the boulevard to complement the five thousand other shade trees called for in the original design of the boulevard. For years thereafter, local Boy Scouts prepared the street for the parade by polishing the small grill-work fences that surrounded each sapling and shining the bronze tablets attached to the slender trunks, each of which commemorated a fallen lo-cal soldier.

Marchers returned to the boulevard the following month for another cherished Bronx institution, the annual Borough Day parade, an event that flourished from 1912 to 1928. In 1921, some fifteen thousand men and women and ten thousand schoolchildren observed the holiday accompa-nied by half a hundred marching bands, ten tractor-drawn siege guns, au-tomobiles bearing members of the Grand Army of the Republic—a score of quavering Civil War veterans who had fought with the Union forces—and a contingent of students from Evander Childs High School carrying placards demanding to know why they didn't have an athletic field. By 1927, marchers had swelled to thirty thousand, a dozen Civil War vets were part of the procession, the crowd numbered nearly two hundred thousand, and participants included a score of hometown girls—Miss Highbridge, Miss University Heights, Miss Bedford Park—vying for the coveted title of Miss Bronx as they crept along the boulevard atop crêpe-paper-drenched floats.

Parades were also a neighborhood staple during the Second World War, although the war at home involved far more sobering rituals than march-ing to the music of brass bands. Of the husbands, fathers, sons, and broth-ers stationed in Europe and the South Pacific, some came home maimed and some did not come home at all; everyone knew who the local Gold Star mothers were. The Kingsbridge Armory, the redbrick castle on Jerome Avenue that was said to possess the world's biggest drill hall—the size of four football fields—brooded over the neighborhood. With its towers and crenellated parapets, the Romanesque Revival building resembled noth-ing so much as a medieval fortress, a constant reminder of the comings and goings of soldiers.

Although the fighting was taking place first on the other side of the ocean and then on the other side of the world, New York City was widely assumed to be a prime target for German bombs, and both physically and psychologically the West Bronx was not many miles from the heart of Manhattan. Every night, air-raid wardens wearing white armbands and matching helmets emblazoned with large triangles patrolled the darkened streets to make sure that all the lights had been turned out, that even the Sabbath candles had been extinguished.

The borough felt its share of wartime casualties; the official listing of Bronx men who lost their lives in the conflict runs to pages and pages of closely typed documents. During those years, parades took on a somber air, and the mood was echoed by other rituals, among them the sale of war bonds outside the lavish façade of Loew's Paradise and the sound of Franklin Roosevelt's fireside chats wafting from every open window. Almost overnight, the streets were drained of men of fighting age. On rooftops, women whose husbands were fighting overseas gathered to read snatches of love letters that had been written thousands of miles away.

So many local families had roots and sometimes relatives in the lands where the war was being fought that the conflict felt unexpectedly near, and Hitler's final solution hardly an abstraction. Long before the larger world learned of the six million Jews who had perished at German hands, large numbers of West Bronx families began hearing fragments of horror stories in letters from abroad. The mother of Bronx-born novelist Jerome Charyn—the dark lady from Belorusse, as he calls her in a memoir with that title—was hardly the only local Jew who spent those years awaiting a letter from a beloved relative perhaps already dead in the conflict.

≡ • ≡

The qualities that made the boulevard a natural route for all those patriotic parades also rendered it irresistible to politicians, that is to say Democrats, a reflection of the iron hand with which party boss Ed Flynn ran the local Bronx organization for so many years and the borough's almost Pavlovian inclination to vote straight tickets. When Franklin Roosevelt made a campaign stop along the Grand Concourse on October 28, 1940, in pursuit of an unprecedented third term as the nation's chief executive, thousands lined the streets to watch his motorcade head north to a reception at Fordham University. President Harry Truman, whose bid to occupy the White House in his own right was helped immeasurably by Boss Flynn's efforts on his behalf, came campaigning in 1948, and one autumn day in 1960, another beloved Democrat drew crowds as rapturous as those that had turned out for FDR.

On November 5 of that year, just three days before the election that put the first Roman Catholic in the White House, a motorcade for John Kennedy swept up the boulevard en route to a rally at Fordham Road and the Grand Concourse, near Loew's Paradise. As the candidate swam into view atop an open convertible, the hysteria reached such a pitch that supporters waving signs bearing such slogans as "The Home of the Knishes Thinks Jack Is Delicious" were nudged aside by women dashing out of beauty parlors, towels over their damp curls, hoping for a glimpse of the hatless, movie-star-handsome candidate. "All I can remember," recalled Rosemary Rogers, a local girl from an Irish-Catholic family who was in grammar school at the time, "is the color of his hair—copper—and being awed that someone with a full head of hair, unlike Ike, might actually be elected president."

"It was a great place to campaign," recalled Herman Badillo, the former Bronx borough president and congressman who trolled the boulevard in search of votes during the 1960s and 1970s. "When I was running for borough president, I shook hands with five thousand people every day—nine thousand on Saturdays and Sundays. I'd stand in front of Alexander's on a sound truck from ten in the morning until four in the afternoon." Someone had the bright idea of making the Brillo scouring pad the candidate's symbol—"Brillo from Badillo; help me clean up the Bronx"—and prospective voters were so enamored of the little soap pads handed out at rallies that they used to line up, take their free sample, then return to the end of the line to collect a second one. The candidate, an enthusiastic runner, also whipped up crowds by jogging up and down the boulevard.

Robert Abrams, Badillo's successor as borough president, did not flood the Bronx with cleaning supplies, nor did he jog. Still, his campaign pursued one original strategy in its effort to win local hearts and minds. "I'm Bob Abrams; have a fortune cookie," the candidate repeated over and over as he handed out cookies to the elderly men and women sunning themselves on the benches of Joyce Kilmer and Franz Sigel parks. To the surprise of no one, the slip of paper inside advised recipients that "the Bronx's good fortune will be Bob Abrams for borough president."

≡ • ≡

When people speak of life along the Grand Concourse, especially during the decades that the street was the center of what they proudly called the Golden Ghetto, these civic, religious, patriotic, and celebratory dramas play a vital role in their memories. Yet an equally powerful shadow was cast by the buildings along the boulevard, not only the Art Deco apartment houses that contributed so mightily to the street's allure but also resonant public places that were destinations in themselves and often of considerable architectural distinction in their own right. The boulevard was home to the borough's splashiest movie theater, its leading hotel, several of its most prestigious synagogues, three beloved parks, a pair of massive monuments born of the Great Depression, a highly peculiar charitable institution, and the gateway to the Bronx's premier shopping district, a few steps from an unassuming cafeteria where one of the greatest criminals in the history of sports learned his trade.

Just west of the boulevard was the home of the most successful professional sports team on the planet, the ballpark that was one of the most famous places in the country. At the northern tip of the Grand Concourse sat Mosholu Parkway, a place where some of America's most celebrated names in fashion and entertainment took the first steps toward what became their brilliant careers. So great was the boulevard's luster that it spilled onto buildings off to either side, notably the public high schools that educated generations of West Bronx girls and boys—DeWitt Clinton, Bronx Science, Walton, and especially Taft, alma mater of Stanley Kubrick and Eydie Gorme—"the Concourse's own high school," as Jerome Charyn always referred to the schoolhouse whose history was so intimately threaded into that of the neighborhood.

The major public and private institutions along the Grand Concourse, the bulk of which arrived between the two world wars, defined the nature of boulevard life as powerfully as the broad contours of the street itself. These buildings and monuments insinuated themselves into collective memory in complex and powerful ways; ordinary and extraordinary lives played out within their walls. If you traveled from south to north, stopping at each of these locations, you would come away with a vivid sense of their contributions to the boulevard's moods and ambiance.

Recollections about these places are occasionally dismissed as mere nostalgia, the implication being that the memories, having grown rosier and hazier over time, are at best unreliable and at worst inaccurate. Sometimes, yes. But these recollections have their own reality, and more important, they offer the only way to understand how people who lived through these years in this place perceived that experience and how it shaped their thoughts and their lives.

⇒ • ⇐

Other than the main post office in Manhattan, if any building elevated the prosaic act of buying a stamp or mailing a letter into an almost religious activity, it was the General Post Office on the Grand Concourse at 149th Street, near the boulevard's southern tip, after the Ben Shahn murals arrived in 1939.

The thirteen panels were inspired by Walt Whitman's poem "I Hear America Singing," and like many art works commissioned by the federal government during the Great Depression, they did not come into being without incident. In the largest image, the bearded poet points to a black-board bearing a passage from one of his verses, urging the world "to recast poems, churches, art; (Recast, maybe discard them, end them—Maybe their work is done, who knows)"—at least those are the words Shahn would have chosen had not a Fordham University ethics professor gotten wind of Shahn's intentions and denounced the passage as "an insult to all religiously minded men and to Christianity."

The priest demanded that plans to inscribe the offending words be scrapped, and to the surprise of those familiar with Shahn's politics, the artist readily complied. He knew from sad personal experience who generally emerged as the loser when art and the state collided, even in a democracy; he had worked with the Mexican artist Diego Rivera on the celebrated *Man at the Crossroads* mural at Rockefeller Center, smashed into powder because Nelson Rockefeller was offended by an image of the Russian revolutionary Vladimir Lenin. And so Shahn agreed to substitute a few of the poet's less incendiary lines, words he hoped would prove more acceptable to the city's powerful Roman Catholic hierarchy.

When he and his wife, Bernarda Bryson Shahn, submitted their proposal for images drawn from Whitman's ode to the American worker, the artist said his aim was to show the men and women of the Bronx, "those people who are as provincial as only city people can be," something about America outside New York. And indeed, the worlds he conjured would have been a revelation to his city-bred audience. These monumental rust-toned images, which dominate the ground-floor lobby of the chaste gray-brick building, its walls pricked with large arched windows, depict workers harvesting wheat, picking cotton, and toiling in steel mills and textile factories far from the heart of the metropolis. Their heads are bent, and their outsize hands and forearms seemed to have swelled to allow the performance of heroic acts.

The eloquence of these figures is enhanced by the setting, a cavernous space flooded half the day with sunshine. "What I remember most," the artist's widow told a reporter years later, "was how the light streamed in through those high arched windows and filled that beautiful hall."

⇒ • ⇐

Very different political imagery was on view in another Depression baby, the Bronx County Courthouse nine blocks to the north. The ten-story mountain of bronze and granite, which also housed the borough's administrative offices, was even more imposing than the post office, rearing up from a sweeping, 125,000-square-foot terrace and reached by a ziggurat of great stone steps.

Mayor Fiorello LaGuardia, who conducted the city's official business in the courthouse during the three-day-long opening festivities, called the place "a golden fortress," and the lawyers and judges and civil servants who trudged up and down the broad steps and roamed the marble corridors seemed dots in a vast landscape, dwarfed especially by the eight monumental sculptural groupings on the terrace. The sculptures, carved from slabs of pink Georgian marble, bore such grandiose titles as *The Majesty of Law* and *Loyalty, Valor, and Sacrifice*, the meanings of which mystified even some architecture critics, one of whom dismissed the rosy concoctions as "allegorical-pompous" and added, "Only the sculptor would know for sure that on the east side of the 158th Street entrance, a somewhat Hellenistic lad with right hand held up in a Marxist-looking clenched fist, and left hand holding a book, flanked by two scantily clad ladies, represented *The Effects of Good Administration*."

Despite the flashes of nudity, nothing about these self-important monuments to civic virtue was the least bit erotic. The neighborhood did, however, have its own pin-up girls, thanks to the Lorelei fountain, which had arrived in the Bronx with such drama in the late nineteenth century. The statue now sat in Joyce Kilmer Park, the seventeen-acre expanse of lawns and pathways just opposite the courthouse that every spring was a sea of red tulips. The park had been named after the poet who lost his life in the First World War, and a few lines of his poem "Roofs"—"But I'm glad to turn from the open road and the starlight on my face, and to leave the splendor of out-of-doors for a human dwelling place"—are inscribed beneath an Art Deco mural in the lobby of 910 Grand Concourse. Though Lorelei herself was discreetly clothed in classical drapery, the mermaids nestled at her feet were lusciously naked, their cold marble breasts temptingly exposed for the delectation of every prepubescent boy in the West Bronx.

And the mermaids were chaste compared with the sort of thing that went on in Franz Sigel Park, the shadowy wilderness south of the courthouse. Screams sometimes drifted from deep within its foliage, and breathless stories were told about people who disappeared into its murky darkness, not to be seen or heard from again. Yet who was to say what really went on there? The park's interior was shielded from the street by virtue of its location atop a steep outcropping and its thick blanket of trees and shrubbery. It was easy to concoct stories about a place so hidden from public view.

⇒ • ⇐

Bronx-born Sam Goodman, age three, in front of the Lorelei fountain in Joyce Kilmer Park in 1955. The ornate marble sculpture arrived in the Bronx in 1899, and for most of its time in the borough the work enjoyed a happy life. (Private collection)

Accounts of the creation of Yankee Stadium invariably begin with the birth in 1895 of an incorrigible boy from Baltimore, such a trial to his parents that at the age of seven he was shipped off to St. Mary's Industrial School for Boys, an institution that was part orphanage, part reform school. Despite an oddly built body in which a muscular torso was balanced atop spindly legs and what one sportswriter later described as "debutante" ankles, the boy was also, and from early on, a spectacular baseball player. His name, of course, was George Herman Ruth.

In 1914, when Ruth was nineteen, he was bought by the Boston Red Sox, who five and a half years later sold him to the New York Yankees in what quickly achieved renown as the most wrongheaded transaction in the history of sports. That year, 1920, Babe Ruth hit an astounding fifty-four home runs and helped draw a record one million fans to the Polo Grounds in northern Manhattan, where the Yankees played as tenants of the field's owners, the New York Giants. In 1921, Ruth helped the Yankees win their first American League pennant and attract another million fans to the ballpark; that year he also led both major leagues with fifty-nine home runs. "They all flock to him," Yankee manager Miller Huggins told reporters in describing his star player, because fans like "the fellow who carries the wallop."

Headline writers tumbled over one another in an effort to do justice to Ruth's remarkable achievements—he was the Sultan of Swat, the Wazir of Wham, the Caliph of Clout, the Maharajah of Mash. Most of all, this player who was as colorful off the field as on it had seemingly overnight transformed baseball from a pitcher's game to a slugger's, a sport in which all eyes were on a man with a bat poised to send a small ball sailing into the sky.

The team needed a property big enough to accommodate the hordes of people who wanted to see Ruth play, and they settled on a site at River Avenue and 161st Street, just across the Harlem River from the Polo Grounds and a few blocks west of the Grand Concourse. In the mid-nineteenth century, the area had been immortalized in a popular Currier and Ives print, one that depicted a lovely pastoral scene dominated by the High Bridge, the Roman-style aqueduct built to transport water from upstate to New York City.

The area had grown steadily less pastoral since Currier and Ives's day; it was home to, among other intrusions, a sawmill, which sat on 11.6 acres belonging to the Astor family. In 1921, Yankees owners Jacob Ruppert Jr. and Tillinghast Huston purchased the Astor property, and on May 6, 1922, shortly after the start of Ruth's third season in New York, laborers began work on a structure that eventually cost two and a half million dollars and would be the nation's largest ballpark.

The impressive numbers that charted the ballpark's creation were the stuff of myth. According to the *New York Times*, some 116,000 square feet of sod were imported from Long Island to cover the diamond and the outfield, and building materials included 30,000 cubic yards of concrete, 3,500 tons of steel, two million board feet of lumber for the bleachers, 600,000 linear feet of lumber for the grandstand seats, 500 tons of iron, and four miles of piping for railings in the boxes, reserved seats, and bleachers. Other accounts cited additional outsized numbers: According

Opening day at Yankee Stadium, on April 18, 1923. The stadium on 161st Street, down the hill from the Grand Concourse, was the country's most famous ballpark, home to the most celebrated professional sports team in the nation. (Bronx County Historical Society)

to the catalogue for an exhibition sponsored half a century later by the Bronx County Historical Society, the grandstand seats required 135,000 individual steel casings, with 400,000 pieces of maple secured to the casings by more than a million brass screws. To bolt the seats to the concrete, more than 90,000 holes were drilled, and to erect the bleachers, 950,000 board feet of Pacific Coast fir were brought via the Panama Canal.

Although these dazzling and sometimes contradictory figures could hardly have all been accurate, the hyperbole reflected the fact that the stadium was not simply a shrine to baseball but a mighty symbol of an optimistic moment in the nation's history. "The stadium can be seen for miles," wrote Fred Lieb, the *New York Evening Telegram* sportswriter who coined the phrase "The House That Ruth Built," "as its triple decks grand stand majestically rises from the banks of the Harlem."

⇒ • ⇐

Opening day was April 18, 1923. When the admission gates opened at 9:30 a.m., for a game scheduled to start six hours later, more than ten thousand people were already waiting in the rain, in a line that extended from the stadium up the hill to the Grand Concourse. Judging by the license plates on the cars in the parking lot, fans had arrived from twenty-five states.

Although the stadium's official capacity was 62,000 seats, an estimated 74,200 fans, many of whom would stand four and five abreast in the grandstand, surged into the park for the opening game. Outside the stadium, the *New York Times* reported, "flattened against doors that had long since closed, were 25,000 more fans, who finally turned around and went home, convinced that baseball parks are not nearly as big as they should be."

Governor Al Smith threw out the ceremonial first pitch, and in the third inning, with the score 0–0 and two men on base, Ruth stepped up to the plate. Generations of sportswriters have vied to capture the drama of the moment that followed; the legendary journalist Grantland Rice, writing in the *New York Tribune*, offered one of the more memorable accounts. "A white streak left Babe Ruth's 52 ounce bludgeon in the third inning of yesterday's opening game at the Yankee Stadium," Rice wrote. "On a low line it sailed, like a silver flame, through the gray, bleak April shadows, and into the right field bleachers, while the great slugger started on his jog around the towpaths for his first home run of the year."

That three-run homer, the first in the new stadium, helped the Yankees to a 4–1 victory over the Red Sox. After the game, as Ruth headed for the clubhouse, he was set upon by two mobs of small boys, one from the right-field stands, the other from the left—at least a thousand youngsters altogether. "They caught and surrounded him," the *Bronx Home News* reported, "and the 'Babe' walked to the door of the runway, flanked by shouting hundreds, who flung their caps in the air and cheered themselves hoarse. 'Babe' had smashed out a home run and once more he reigns in the kingdom of boyish Bronx minds"—a state of mind that the writer Neil Sullivan, retelling the stadium's history three-quarters of a century later in his book *The Diamond in the Bronx*, described as "an extraordinary realm of hope and imagination in which the most common dream was no doubt each boy's determination to someday replace the Babe himself."

Ruth went on to hit forty more homers that year, forty-six the next. And so began the team's glory years, a mostly unbroken stream of breathtaking, heart-stopping performances that bathed the stadium in glory and its neighborhood in reflected luster. In the 1920s, when Ruth shared some of the limelight with the almost equally talented but less celebrated Lou Gehrig, a ferocious batting lineup earned the team the nickname Murderers' Row; in the eyes of some students of the sport, the 1927 Yankees,

A Yankees victory dinner, one of countless celebratory events held at the
Concourse Plaza Hotel. (Bronx County Historical Society)

who racked up 110 victories, with Ruth hitting a record sixty home runs,
was possibly the most formidable team in baseball history.

Ruth retired in 1935. But the Bronx Bombers, whose numbers went
on to include Phil Rizzuto, Yogi Berra, and Joe DiMaggio, continued to
amaze. In one notable winning streak, which began in June 1947 with four
games against Boston, the strongest team in the league, the Yankees scored
forty runs to the Red Sox's five. "I cannot think of one contending team so
trampling another and so humiliating them," Roger Angell wrote of those
games more than half a century later. On June 20, the Yankees took first
place, and they never gave it up. The season climaxed in what Red Barber
called "the greatest World Series ever played," in which the Yankees beat
the Brooklyn Dodgers in seven games.

In the 1950s, the decade dominated by Mickey Mantle, there were three
teams in town, and a New York club played in every World Series from 1949
through 1958, although *Damn Yankees*, the title of a popular Broadway musical, made clear which was the team to beat. In 1961, Roger Maris surpassed
Ruth with a record sixty-one home runs. In the team's years in the stadium,
the Yankees won thirty-nine pennants and twenty-six World Series.

The schemes by which local youngsters sneaked into the ballpark were almost as much the stuff of legend as the statistics about the stadium's construction. A boy named Bobby Parrella used to enter the stadium under the cover of busloads of Catholic orphans. He and his friends waited for the buses to pull in, then slipped into the crowd, making sure to take one of the bag lunches being handed out. "You had to be careful not to get taken back with them," Parrella, who went on to be a columnist for the *New York Herald Tribune*, recalled years later. "But we were never questioned. We looked like orphans, I guess. We ate those sandwiches hungrily."

Baseball was not the only attraction. On June 22, 1938, nearly seventy thousand spectators converged on Yankee Stadium for one of the most widely publicized boxing matches in history, a contest in which Joe Louis, the twenty-four-year-old black American world heavyweight champion, defeated thirty-two-year-old Max Schmeling, the greatest athlete of Nazi Germany. Football was also played in the stadium, and in later years, popes recited Mass there, and evangelists trolled for souls.

Baseball, however, was the ballpark's heart. Clocks and calendars moved according to Yankee time. From spring to fall, the neighborhood smelled like a giant hot dog. There was something almost eternal about the stadium's configuration, as if the place were a bizarre relic from the ancient past, as author Laura Shaine Cunningham, who lived almost literally in its shadow during the 1950s, suspected. "Golden, softly rounded, the old stadium had a Biblical look," Cunningham wrote in her memoir *Sleeping Arrangements*. "I assumed it had been standing in 161st Street since before Christ." And like virtually everyone in the neighborhood, she and her friends succumbed to the Yankees' spell:

> Every baseball season . . . I look forward to the annual visitation of the godlike Yankees, who march across Grand Concourse. . . . I can't imagine a more momentous event than the arrival of the Yankees at the Concourse Plaza. The Yankees—sturdy, handsome—seem to arrive to save us from some fate. . . .
>
> Am I the only one who thinks Mickey Mantle may be more than mortal? His alliterative name, his batting average, the very word "Mantle"—all seem to imply that he may be more than human. I regard Mickey Mantle as enjoying some in-between status—part human, part deity, all Yankee. (Is this wishful memory, or did they really wash the street before he crossed it?)

If they didn't, maybe they should have. The cleansing would have been a fitting acknowledgment of how much the team meant to the neighborhood, how large it loomed in local hearts and minds. Though the action took place on the field, the stadium's luster resonated mightily throughout the West Bronx. In October, World Series time, families camped out under the El on River Avenue, near the entrance to the bleachers, waiting for tickets to go on sale. Throughout the season, local children perched on the rocks of Joyce Kilmer Park and listened to broadcasts of the game from transistor radios held close to one ear, the sound of the announcer's voice echoing the cheering down the hill.

And after performing so memorably on the field, the team's strong and confident young players wandered the neighborhood's very streets. On game nights, Babe Ruth, Mantle, and others hung their hats at the Concourse Plaza Hotel, as did many members of visiting teams. The Yankees celebrated their seemingly nonstop victories in the hotel's Grand Ballroom and could often be seen passing through the lobby, generally with knots of small boys in pursuit.

⇒ • ⇐

The hotel, the redbrick and limestone building at 161st Street whose façade was accented with wreaths and urns and other classical details, is much on the mind of Aggie Hurley, the frumpy, social-climbing Bronx mother played by Bette Davis in the 1956 film *The Catered Affair*. Davis's character spends prodigious energy arranging an elaborate wedding breakfast for her newly engaged daughter, a dewy Debbie Reynolds, despite the fact that the girl could care less about a big affair, and the cost would wipe out her cabbie father (Ernest Borgnine), who has been saving for years to buy his own taxi.

Paddy Chayefsky, who wrote the television play on which *The Catered Affair* was based, had just a year earlier depicted Bronx working-class life in far more nuanced fashion in the Oscar-winning film *Marty*, his mournful portrait of a restless young butcher. Yet the ghost of Marty flickers through Borgnine's performance, and Davis's yearning for the lavish wedding she never had—"one fine thing to remember when the bad days come"—seems sadly genuine. Although the catered affair of the title does not take place—all the audience sees is Borgnine pulling his taxi up to the hotel's famous canopied entrance, checking out the Grand Ballroom, and haggling with the banquet manager over the price of flowers and how many guests can be crammed into a single limousine—with the choice

The elegant Concourse Plaza Hotel, a residential establishment that opened in 1923 on 161st Street and went on to become the borough's most prestigious hotel, was only "thirty minutes from Wall Street," as a promotional brochure boasted. (Bronx County Historical Society)

of the Concourse Plaza as the ultimate setting for a showy Bronx affair that would dazzle the neighbors and offer the memories of a lifetime, the movie got one crucial detail exactly right.

The hotel had appeared on the scene thirty-three years earlier, a moment at which optimism about the borough's prospects was even headier than it had been when the Bronx was poised to become part of Greater New York. A group of enterprising local businessmen calling themselves the Bronx Boosters had met in April 1921 to discuss construction of what they described as a "high-type apartment hotel," and from there things moved fast. Workmen broke ground the following March, and by October 1, 1923, the Concourse Plaza Hotel was ready to welcome its first residents.

The arrival of the ten-story, $2.5 million hotel seemed a harbinger of the good fortune poised to envelop the entire borough. As Governor Al Smith, guest of honor at the opening banquet, informed the nearly one thousand people who had gathered in the Grand Ballroom, "The Bronx is the most striking example of urban development in the United States," and for decades his words were quoted as a reminder of the borough's potential.

Although the hotel had much to offer, one of its great assets was its incomparable setting, "at the broad-spreading entrance to one of New York's handsomest boulevards," as a promotional brochure entitled "30 Minutes from Wall Street" described the site. As for the vistas this marvelous location afforded,

> Through his window in an upper story the Concourse Plaza tenant views the world's metropolis from an airy and peaceful height. Looking to the south, one has the illusion that the Concourse Plaza is on a plane with the Metropolitan Tower and Woolworth building. To the west one sees the Palisades of the Hudson and to the east the waters of Long Island Sound. To the north stretches the beautiful green parkway of the Grand Concourse. At night the myriad lights of the city are a fascinating sight.

The description made it sound as if no one ever wanted to leave, and for years that was literally the case. Despite its name, the Concourse Plaza was not actually a hotel but a high-class apartment house, one that nonetheless offered all the amenities a luxury tenant could desire, in a setting presided over by a manager who had polished his skills at the Ritz in Paris. Its 160 apartments of up to four bedrooms each came equipped with kitchens and dining rooms, china closets, foldout Murphy beds, Kilmoth cedar closets "permitting storage of furs," and according to information provided to guests, the services of "chambermaid, bathroom maid, charwoman, houseman, porter and bellboy, vacuum cleaning, window washing, etc." Breakfast was served in the grill, and guests could buy "the best quality of provisions" at an on-site commissary or dine at the terrace restaurant that overlooked the boulevard, a carpeted, softly lighted place where an orchestra played and furnishings were in the Adam style even down to the silverware. For more formal events, a French chef prepared such delicacies as turtle soup and lobster thermidor, served with endless bottles of champagne.

In the richly paneled lobby, accented with fat white pillars, marble floors, and potted plants, local officials and Democratic Party bosses held perpetual court, collecting political debts, dispensing largesse, and once in a while conducting a spot of real business, generally of a dubious nature; the suites and offices they maintained at the hotel were sometimes literally a home away from home. These individuals also attended an endless stream of civic and political events, which, like so many of the hotel's dramas, played out in its ballrooms—three floors' worth of crimson carpet and gilt wallpaper with such names as the Wedgwood Room, the Gold Room, the Crystal Room, and especially the Grand Ballroom, seventy-five feet wide and a hundred feet long, with twenty-eight-foot ceilings. So vast was this space that it felt crowded only when occupied by more than two thousand dancers or one thousand diners. Hundreds of lights from crystal chandeliers illuminated its soft old-rose and silver décor, and marble trimming and glossy wainscoting added to the sense of richness. Less energetic guests watched the proceedings from velvet-trimmed banquettes in the balcony—a broad promenade banded by a gilded railing—like characters from a Jane Austen novel observing the merrymaking at the village ball.

Local lore had it that every Bronx boy called to the Torah celebrated the event in one of these gorgeous rooms, that every Bronx bride headed directly to the hotel just moments after she and her new husband tied the knot. And although each wedding was pronounced more sumptuous than the last and each bride lovelier than her predecessor, in certain circles it was agreed that Betty Kanganis, whose father ran the florist shop at Kingsbridge Road and Jerome Avenue, had the most exquisite flowers.

Betty's future husband, Nick Raptis, worked in the shop. They had met in 1947 at the flower dance at the Pennsylvania Hotel in Manhattan, and as Betty recalled more than half a century later, their love affair was the talk of the Greek American community of the West Bronx, mostly because Betty was just a sophomore at Walton High School, her sweet-sixteen party still a year away. Two years later, after a courtship conducted under the watchful eye of Betty's grandmother, the marriage took place on February 27, 1949. The bride was seventeen, the groom nearly twenty-five.

Father Basil Efthimiou performed the Greek Orthodox ceremony at the Cathedral of Holy Trinity in Manhattan. Betty wore an ivory satin gown from Jay Thorpe, one of the city's better department stores, hand-trimmed with lace and seed pearls, along with a matching beaded crown and her glossy black braids piled high on her head, giving her the look of a young Greek princess. On the third finger of her left hand was a four-hundred-dollar diamond-and-gold engagement ring that her husband-to-be had emptied his bank account to buy from an Armenian jeweler in Washington Heights. In her arms was a bouquet of orchids and lilies of the valley made by her father.

After the ceremony, the 320 guests headed up to the Concourse Plaza to feast on a wedding cake nearly half as tall as the bride and to dance to the music of the George Kravas Orchestra, the city's reigning Greek musicians. Thanks to the bride's father, the Grand Ballroom had been transformed into a summer garden. Deep swags of greenery draped the main table and hung from the balcony, and huge vases of orchids and red and pink camellias stood in every corner. In one of the photographs in the couple's white satin wedding album, the newlyweds kiss beneath a lacy palm tree, as if they had momentarily escaped from the West Bronx into a tropical Forest of Arden.

The photographs were taken by an Armenian refugee named Kourken Hovsepian—Mr. Kirk was his professional name—who for more than half a century operated a studio on Kingsbridge Road between Morris and Creston avenues, a few blocks west of the Grand Concourse. His studio was located just above Spotless Cleaners, along a commercial strip that included such other icons of midcentury urban life as Sol and Moe's candy store, Brownstein's hardware, Star Drug, G and R Jewish baker, the notions shop, the Jewish appetizer store, the kosher butcher, the Irish bars. "He'd tell people it was between the Concourse and Jerome Avenue," recalled Mr. Kirk's daughter, Mimi Vang Olsen, an artist who lives in Greenwich Village and is the custodian of her father's photographs. "He thought it sounded better."

Mr. Kirk was one of the main chroniclers of the occasions that defined the West Bronx. In his images of thousands of celebratory events—Jewish bar mitzvahs, Greek weddings, Irish christenings, Italian confirmations—a lost world swims momentarily into focus. Starry-eyed brides emerge triumphantly from a sea of attendants, their enormous trains swirling about them like giant organdy fishtails. Endless processions of bar mitzvah boys, barely daring to move in their brand-new suits and crisp white shirts, smile bravely for the camera. The images are so detailed, so evocative, that you can almost hear the buoyant tarantellas, the cantor's drone, the rustle of taffeta, the clink of champagne glasses. You long to sink your fingers into the Persian lamb coats. You can almost smell the Shalimar.

⇒ • ⇐

Unlike the ballpark and the hotel, the most elegant structure in the West Bronx did not thread itself into the fabric of neighborhood life. The Andrew Freedman Home stood aloof on a little hill on the Grand Concourse between 166th Street and McClellan Street, an Italian Renaissance palace of soft gray and yellow limestone that occupied an entire oversized city block and looked down almost literally on its less pedigreed neighbors.

In its lofty indifference to its surroundings, the Andrew Freedman Home seemed the perfect reflection of the character of the man whose

Betty Kanganis, whose father ran a florist shop on Kingsbridge Road, with her new husband, Nick Raptis, dancing in a garlanded ballroom at the Concourse Plaza Hotel at the couple's wedding in 1949. The photographer was an Armenian immigrant named Kourken Hovsepian—Mr. Kirk to generations of West Bronx families. (Courtesy of Mimi Vang Olsen)

name it bore. At a moment when New York was being flooded with impoverished immigrants from Eastern Europe, this son of German Jews established an institution not to serve his coreligionists and alleviate their wretched living conditions but to allow carefully selected members of the city's formerly rich to live out their days in a style to which they had become accustomed during their plummier years. Although Freedman was not atypical of his set in turning his back on the Jews that came after him, his ostentatious indifference to these newcomers made him hard to admire, even by the standards of the day.

Andrew Freedman was born in New York City in 1860, attended the City University of New York, and then embarked on a career in real estate, a field in which he proved extremely successful, albeit, as the *Dictionary of American Biography* noted enigmatically, "by ways that are no longer traceable." A loyal Tammany man, Freedman played a key role in the construction of the city's original subway line, which was built by the Interborough Rapid Transit Company, credited with bringing together the project's main architects, contractor John McDonald and financier August Belmont. For this achievement, Freedman is sometimes called "the father of the New York subway." Yet here, too, his role is mysterious, so much so that at the time of his death in 1915, the state legislature was poised to investigate his complicated relationships with the companies involved.

Especially after Freedman founded his namesake institution, he was lauded as one of New York's illustrious personages, a man whose skill and vision helped shape the modern city. But judging by newspaper accounts of his day-to-day activities, the real Andrew Freedman was an ill-tempered, hot-headed man, despised by many, loved by few, and apparently incapable of staying out of trouble, generally of his own making. His personality seemed writ large in his physiognomy; even in his most winning photographs, his black eyes glare angrily from beneath thick, caterpillar-like eyebrows, and his mouth, almost hidden by a ferocious mustache, seems set in a perpetual scowl.

Freedman's cantankerous nature manifested itself most spectacularly during his eight-year reign as owner of the New York Baseball Club, the team later known as the New York Giants. He hired and fired a dozen managers, launched twenty libel suits in a single year, and, during one game, when a former member of the New York club announced to the crowd, "I'm glad that I don't work for a Sheeny any more," charged down the aisles accompanied by a cadre of truncheon-wielding private policemen, intent on throwing the player off the field. "Freedman's irascible personality,

Andrew Freedman, the enigmatic nineteenth-century New York financier who used his vast fortune to create what the *New York Times* called "an ex-rich man's poorhouse." (Private collection)

quick temper, and aggressiveness had him in constant trouble," one base-ball historian wrote of his tenure. "[He] kept the league in constant tur-moil and made the *New York World's* observation that he had 'an astonish-ing ability for making enemies' seem like a gross understatement."

Despite a fortune estimated at his death at five million dollars, Freed-man's personal life seemed hardly more satisfying. It was said that after a broken engagement he retired into a shell, living out his life as a testy bachelor, the only important women in his life his mother and his sister, Isabella.

As to what shaped Freedman's obsession with the plight of the formerly rich, one can merely speculate. The explanation most often cited is that his parents, once successful, lost their money when their son was still young. The Panic of 1907, the brief but devastating collapse of the stock market, must have been terrifying. And having amassed his wealth by his own hand, Freedman presumably understood better than most how fortunes could evaporate overnight. In any case, apparently appalled at the idea of a rich person having to live out his final years among the chronically des-titute, a group Freedman never much cared for in the first place, he wrote an extraordinary will.

The document provided for "the establishment of a nonsectarian home where gentlefolk, people of culture and former wealth, who had once been in good circumstances, but who, by reasons of adverse fortune, had be-come poor, could live in the comfort to which they had been accustomed." Although it was assumed that guests would have sufficient resources to cover their formal attire, their cigars, and their bridge lessons, there was to be no charge for their care and accommodations, and what was then the astounding sum of seven million dollars was earmarked to make this grand scheme a reality.

Nine years after Freedman's death, in May 1924, the home that bore his name opened its doors. "An ex-rich man's poorhouse, stripped of every suspicion of charity," the *New York Times* remarked. "In accordance with the wishes of the founder, the Freedman Home will be . . . operated solely for the care and maintenance of gentle folk of advanced age who were once wealthy and now live in penury, but who have preserved their taste for . . . comforts and refined surroundings."

Isabella Freedman, apparently as big a snob as her brother, echoed these sentiments. "There are homes for people who have always been poor," she told a reporter on the eve of the opening. "There are homes for impover-ished people of the middle class, but there is none for people who were

The Andrew Freedman Home on the Grand Concourse at 166th Street. This unusual charitable institution, its design modeled after the Palazzo Farnese in Rome, allowed selected members of the city's formerly rich to stave off an impoverished old age. (*New York Times*)

once rich, who were accustomed to live in luxury and refinement and who for that reason find poverty at old age doubly hard to bear."

Even Freedman's critics would have been hard-pressed to imagine a more palatial place in which to spend one's twilight years. The building's architects, Joseph Freedlander and Harry Allen Jacobs, had studied at the École des Beaux-Arts in Paris and modeled their creation after the Palazzo Farnese in Rome, one of the city's most majestic structures. The three-story façade, punctuated by rows of arched and pedimented windows, sat behind a wide terrace protected by a handsome balustrade and facing a deeply sloping expanse of landscaped lawn, made more lush over the years as residents planted their favorite trees.

In the seventy-five-foot-long grand salon, where aging stockbrokers dozed in velvet armchairs and overstuffed Georgian sofas, smoking cigars and dreaming of deals brought off in headier years, the furnishings included Oriental rugs, baby grand pianos, and tables topped by antique bronzes. Chandeliers of dull bronze hung low from the fifteen-foot-high ceiling, and a log fire crackled in the marble fireplace. In the oak-paneled library, leather-bound classics from another age—*Drums along the Mo-hawk, The Mill on the Floss*—filled floor-to-ceiling shelves, the top levels reached by small ladders. Amenities included a billiards room and a hundred well-appointment bedrooms, each decorated in an individual color scheme so residents visiting one another's quarters would not get bored by the décor.

Every night, men in black tie and women in flowing dresses dined at tables covered with gold cloths and set with fine china, crystal goblets, and silver forks, there to be attended by white-gloved waiters, some of whom had been recruited from the Cunard Line. The staff of fifty also included a dietician who used to work at Schrafft's, along with a pastry chef reputed to be "a wizard at making centerpieces for special occasions in the shape of miniature houses that light up inside."

Despite the founder's expressed wishes, not to mention a rigorous screening process, the home's early occupants were practitioners of fairly modest occupations—jewelers, dressmakers, nurses, teachers—hardly ex-millionaires. Yet over the years, and as the Great Depression worsened, there arrived a more upscale crowd that included doctors, lawyers, actors, and opera singers. In the spring of 1933, when Geoffrey T. Hellman from the *New Yorker* paid a visit to what the magazine described as a "Bronx Palace," some 129 guests were in residence, among them members of the Daughters of the American Revolution, and complaints were already being lodged about a decline in clientele. A Colonial Dame had actually resigned, the magazine reported, "claiming that the social tone of the membership was falling off."

=== • ===

A young boy named Victor Geller, whose mother operated what he called a shirt hospital on the Grand Concourse near the Freedman Home and who later described his childhood in an affectionate memoir, tells a charming story about one of the home's more celebrated residents, an elderly former general from the Czar's army. Impoverished after the revolution, the ex-military man was anxious to avoid the expense of buying new clothes and so brought his threadbare shirts to the boy's mother to get the cuffs and collars turned for a few cents apiece.

This resourcefulness was understandable; it is hard to picture the Russian aristocrat, with his flowing mustache and Van Dyke beard, making his halting way up the Grand Concourse to Fordham Road to replenish his wardrobe. Still, apart from the neighborhood's wealthiest residents, who patronized the city's better department stores on Fifth Avenue, it was the rare West Bronx housewife who did not pay regular visits to the shopping district that by the 1920s was being touted as the "Forty-second Street of the Bronx." And starting in 1938, vast numbers of these shoppers could be found attacking tables piled high with merchandise and rummaging ferociously through the drawers beneath display shelves at the four-story brick building whose red neon sign informed the shoppers of the world that "Uptown, it's Alexander's."

The chain had been synonymous with the Bronx since 1928, when a businessman named George Farkas opened the first Alexander's, named after his merchant father, at the Hub, the tangle of department stores and other businesses centered on the East Bronx intersection of 149th Street and the Third Avenue El. As population shifted toward the West Bronx, the borough's commercial center of gravity followed, and a decade after the company's founding, Alexander's took over the home of the old Adam-Wertheimer's department store on the northwest corner of the Grand Concourse and Fordham Road at the shopping strip's unofficial gateway.

Alexander's was not the place to find high fashion at bargain-basement prices. For that, New Yorkers descended on Loehmann's, a few blocks to the west. Nevertheless, for shoppers with pretensions to style despite pinched budgets, Alexander's was so legendary a destination that the bargains themselves sometimes took second place to the jousting that was the house sport. A woman who as a young girl made regular pilgrimages to the store with her mother remembers watching, transfixed, as two shoppers nearly came to blows over a marked-down Dacron blouse. At its peak, Alexander's occupied three hundred thousand square feet of space, ten times its original size; at one point it boasted of generating more sales per square foot than any other department store in the world. In an uncanny manner, its intrepid bargain hunters seemed to be heeding the sometimes gnomic admonitions to thrift inscribed on the façade of the Dollar Savings Bank just across the street, among them this advice: "Without economy none can be rich—with it—few will be poor."

⇒ • ⇐

The shopping district at Fordham Road and the Grand Concourse had other claims to fame. John F. Condon, the elderly Bronx schoolteacher who called himself Jafsie and achieved a certain renown as a go-between in the 1932 kidnapping of the infant son of Colonel Charles Lindbergh, spent a considerable amount of time at the restaurant on the Grand Concourse and

188th Street owned by his friend Max Rosenheim. At the same intersection stood a modest establishment with an even more colorful reputation.

Although the Bickford Cafeteria offered little in the way of ambiance or architectural flair, this branch of the popular chain had the distinction of being the favored hangout of Joey Hacken, the neighborhood's popular bookie. A stocky, pudgy-faced man with a perpetual five o'clock shadow, Hacken seemed to have arrived at Bickford's by way of central casting. Though the son of Jewish immigrants, his dark hair and swarthy features gave him a vaguely Italian look, and his workday uniform consisted of a sports shirt and a shabby overcoat, a bundle of newspapers tucked messily under one arm. Friends called him Joe Jalop, a reference to the wreck of a car he drove.

Most days, Hacken could be found just inside the cafeteria's entrance, stationed at the first table on the left and surrounded by a knot of boys from the neighborhood drawn by the promise of action: Who was the favorite for the game that night? Who looked good for the fight? Hacken bought his young acolytes pie and sodas, and they in turn listened avidly as he taught them, largely by example, how to navigate the underside of sports. By 1943, the crowd included a skinny kid not yet in his teens who introduced himself as Jack Molinas.

Jacob Louis Molinas had been born in 1932, the child of a Sephardic Jewish father from Turkey and a mother of Turkish stock. The family lived at 2525 Grand Concourse, a block north of Fordham Road. Upon the arrival of the couple's firstborn son, gold coins were sprinkled into his crib, a traditional gesture intended to ensure a life of good fortune, and the hoped-for blessings were apparent long before the child was out of short pants. A beautiful boy with glossy hair, a radiant smile, and a remarkable brain, he learned to read at the age of three, and his IQ was said to be a genius-level 175. By the time he was twelve, he was five foot ten—in a few years he shot up to six-six—and starting to do amazing things with a basketball in the playground of Creston Junior High School, not far from his apartment.

After the games, Molinas headed over to Hacken's table at Bickford's, there to be tutored about odds, point spreads, and the myriad ways to turn a quick and illegal buck on the backs of unsuspecting ballplayers. The lessons continued at the mob-controlled restaurant Molinas's father owned on Coney Island, a popular racketeers' hangout where the boy's teachers included Izzy the Bug, Frankie the Wop, Shpitz the Galitz, and Shpitz's sidekick, Pork Chop. The owner's son proved an exceptionally apt pupil.

First in high school, then in college, Molinas continued to play brilliant basketball. By his junior year at the city's highly competitive Stuyvesant High School, he was being touted as a future college All-American, and scouts were describing him as the best high school player in the country. As a senior at Columbia University, he was captain and top scorer of the 1952–53 team that won the Ivy League championship that year, an All-American forward named the most outstanding athlete in his graduating class; not until Bill Bradley came along a decade later did the Ivy League produce another such virtuoso in the sport.

But Molinas's real passion was reserved for those activities that took place off the court, activities of which few people were aware. Through all those years, working with his mob-controlled bookie, he threw game after game. He bet regularly against his own team and not only fixed games himself but more insidiously persuaded others to do the same.

In Molinas's senior year of college, the Fort Wayne Pistons came calling, and Molinas, their number-one draft pick, was for a time a National Basketball Association All-Star. Only briefly, however. His criminal past caught up with him midway through his first season with the Indiana team. Molinas was ignominiously led away in handcuffs and in short order banned from the league for life for betting on games illegally.

Yet his banishment from the NBA seemed to feed his fever. A major player in the point-shaving scandals that rocked college basketball in the late 1950s and early 1960s, Molinas masterminded a nationwide gambling ring that involved twenty-seven colleges and almost fifty players, many of whose careers were destroyed. At one point, he and his partners were clearing an impressive fifty thousand dollars a week.

In 1963, Molinas was convicted and sentenced to prison for his activities. By the 1970s, he had ended up in Los Angeles, a producer of X-rated films, and in 1975, at the age of forty-three, he was fatally shot in the head execution style at two in the morning in the backyard of his home in Hollywood Hills. His death, an apparent mob hit involving unpaid debts, occurred a week before he was scheduled to stand trial on pornography charges. In investigating the matter, the police said they were looking into a possible link with the murder of Molinas's former business partner, who had been shot a year earlier.

Molinas would be remembered as probably the greatest fixer of basketball games in the history of the sport. Yet well after his charmed life began unraveling, he remained a hero back in the old neighborhood. "When he walked down the street, all the younger kids would dash through heavy traffic on the Concourse or Fordham Road just so they could walk past Jack and nod hello," the novelist Avery Corman, who grew up on the boulevard, told Molinas's biographer, Charley Rosen. "Even when he was just walking along by himself, it was like he was on a stage."

The question of why such a blessed and brilliant athlete went so spectacularly wrong haunted nearly everyone whose life he touched. Neil Isaacs, a sportswriter who wrote a fictionalized biography of Molinas, suggests that the answer lay in his genes, that the man with the magic touch on the court was "born bent." Molinas's biographer attributes his subject's failings to growing up under the shadow of a mushroom cloud; Jack was thirteen, practicing hook shots in the Creston schoolyard, when the bomb fell on Hiroshima. The man's gentle, almost fatherly bookie was something of a surrogate parent, a far more appealing creature than Molinas's own harsh and unloving father. Maybe Molinas was simply addicted to gambling, a man for whom, as Rosen put it, "playing in a rigged ball game was more exhilarating than playing it straight."

This compelling, almost tragic figure lived a life wreathed in questions. But what the Bronx-born political scientist Marshall Berman found most fascinating about Molinas was the unprepossessing setting where his tortured journey began. "Here he was, probably the greatest crook in the history of sports," Berman said. "And it all started at this little cafeteria on the Grand Concourse."

≡ • ≡

Poe Park, the two-acre triangle on the Grand Concourse at Kingsbridge Road, a few blocks north of Fordham Road, was from its earliest days a lovely place to sit and read a book, to bring an infant in a white wicker baby carriage, or to find a hint of breeze on a hot summer evening. Visitors called Poe Park a "little green jewel" and compared its lush grass to a green velvet carpet.

But the park really started coming into its own in 1925, when a small white bandstand was built at the southern end—"a handsome ornament on this pretty historic spot," remarked the *New York Evening Journal*—and the music started playing. During the 1930s, orchestras performed classical concerts—festive affairs much beloved in a neighborhood whose residents had deep European roots and a passion for serious music—and so thick were the crowds that they often spilled onto the sidewalk. To boost wartime morale, Consolidated Edison sponsored jazz concerts during the 1940s, and the 1950s brought themed evenings: square dancing on Mondays, classical music on Thursdays, and sometimes on weekends, German bands wearing lederhosen. On Wednesday night, dance night, local rock-and-roll groups often stopped by for an evening of doo-wop, among them a sultry-looking teenager from Belmont named Dion DiMucci, a member, or so it was said, of the infamous Bronx gang the Fordham Baldies.

At the northern end of the park stood a white-shingled cottage with a tragic history. The little wood-frame house had been built in 1812 in what was then the village of Fordham. Edgar Allan Poe moved there in 1846, hoping that the fresh country air would be beneficial to his young wife, Virginia, who was dying of tuberculosis. There the poet wrote "Ulalume" and "Annabel Lee," the mournful requiem to his beloved spouse.

Although already famous—"The Raven" had been published the previous spring—Poe was so poor that his mother-in-law, who lived with the couple, used to scour the fields for edible herbs to feed the household. But nothing could be done for Virginia, who died on January 30, 1847, at the age of twenty-six. The poet left the cottage in the summer of 1849, and that October, his mother-in-law received word of the death of a stranger in a Baltimore hospital who used the name "Edgar A. Poe."

In 1913, the city bought the house, moved it to the park from its original location on the other side of Kingsbridge Road, and restored the structure, preserving the bed with the straw mattress in which Virginia Poe had died, along with a mirror, a Bible, a spoon, and the rocker in which the poet sat when he wrote "Annabel Lee." Over the next decades, up to forty thousand people visited the house each year, sometimes as many as a thousand on Saturdays and Sundays. A caretaker during the 1930s named Mrs. Martin Kapp used to soften the details of Poe's story for younger visitors, lest they be unduly influenced by tales of his famous alcoholism. According to a writer for the New Yorker who stopped by the house in 1930, "she always explains to schoolchildren that Poe never drank and was intoxicated by his literary powers." Although possibly dubious about such claims to virtue, visitors were nonetheless invariably struck by how small and dark the house was, how low the ceilings were, how crowded the cottage must have been when three adults lived under its tiny roof, and how unhappy the poet must have been during the years in the Bronx that later brought him such glory.

⇒ • ⇐

No street in the Bronx possessed the mythic trappings of the Grand Concourse. But a little of its luster spilled onto Mosholu Parkway, the strip of greenery at the boulevard's northern tip, and especially to the block-long stretch opposite Public School 80, later Junior High School 80, where in the 1950s a remarkable generation of Bronx children came of age.

In a nostalgic visit to that world nearly three decades later, *New York* magazine christened the boys and girls who hung out along the parkway's iron railing the "Parkway All-Stars." Precocious even then, these lively precursors of the showmen and performers and merchant princes into which they evolved used to gather after school and into the evening, their talent already in evidence. Ralphie Lifshitz, still a few years away from his reincarnation as Ralph Lauren but nonetheless wearing white bucks and English-style jackets with wooden toggles when the other boys favored motorcycle jackets, showed a flair for fashion that had manifested itself in the blue and white satin jackets he designed for the members of his basketball team, the Comets. (Calvin Klein, another local boy with a precocious fashion sense, did not hang out along the railing, but he could be found trailing after his mother up and down the aisles of Loehmann's, analyzing the cut of the designer suits.)

A sharp-featured kid named Robert Klein was gaining a local reputation as a member of the singing group called the Teen Tones. The group also included a girl with buck teeth and braces named Penny Marshall, whose older brother, Garry, would re-create the carefree spirit of Parkway life in his hit television series *Happy Days*. Garry Marshall's best friend, Bernard Gwertzman, one of the few Parkway alumni whose life's work did not involve either amusing people or designing their clothes, went on to become the foreign editor of the *New York Times*.

Gwertzman and Garry Marshall lived at 3235 Grand Concourse, the last apartment house on the boulevard's northern tip. Their hearts, however, were on the parkway, and especially with their baseball team, the Falcons. "The Falcons were a big part of our lives," Gwertzman remembered. "We held a raffle—our parents knew nothing about this, of course—to raise money to buy jackets, which by the way looked very much like mobsters' jackets. There were other teams—the Jaguars, the Sharks—but we were a peaceful little people."

Not all the children of the Parkway flourished. One who skidded badly off the rails was a disaffected loner named Roy Drillich, generally considered an inspiration for Arthur Fonzarelli, the lovable greaser of *Happy Days* made famous by Henry Winkler. Drillich came to a far less happy end. Early on, he became involved with drugs, and in 1980, shortly before the rapturous article about his generation appeared in *New York*, he slashed the throat of his seventy-seven-year-old father and then his own. At the time, both men were living in the house in which Drillich had grown up.

For the most part, though, this neighborhood and dozens like it throughout the borough and beyond were, at least in memory, almost halcyon places. "The soul's geography," the writer and former City University professor Leonard Kriegel, who came of age in the neighborhood during those years, said in describing the nature of life on the streets. Mosholu Parkway was atypical in the number of children who grew up to be major players in Hollywood and on Seventh Avenue. It was, however, utterly typical in the texture and rhythms of one's daily existence, especially for the young. During the relatively unruffled postwar years, and despite the vastness of the city as a whole, enclaves like this one throughout New York, and especially in the outer boroughs, had the texture of small villages.

The radius of one's existence might extend no further than a handful of blocks in either direction, to the schoolyard, the candy store, the friend who lived down the street. One's neighbors and shopkeepers were as intimate a part of one's life as parents and brothers and sisters. The street was a world unto itself, and the pulse of the street, with its seasons, its rituals, its unvarying alliances and hostilities, was the heartbeat of the community.

"An Acre of Seats in a Garden of Dreams"

OF ALL THE MEMORY-DRENCHED spots along the Grand Concourse, the place remembered with the greatest fondness is almost certainly Loew's Paradise, the gorgeous picture palace just south of Fordham Road. Decades after its heyday, the mere mention of its name elicits sighs, even among people who have not laid eyes on its Italian Baroque splendor for more years than they can count. Yet one of the most eloquent tributes to Loew's Paradise takes the form of a grainy, six-minute video made by a young visual artist who was born four decades after the theater's opening in a city more than two thousand miles away and who came to know the theater only after its glory days were long gone.

The artist, who comes from Mexico City, is Pablo Helguera, and in a dreamy work made in 2005 and entitled "Paradise Regained," he reflects on the central position that this aptly named playhouse occupied in the minds and imaginations of midcentury Bronxites. As jerky black-and-white images flicker across the screen and tinny piano music plays in the background, Helguera reminds us that the Paradise "reveals a great deal about a time when the experience of visiting a movie theater had a magical meaning, when one's physical presence in an ideal environment was an integral part of moviegoing." In an elegiac aside, he adds, "At a time when life in the city sometimes hits us with sharp realities, or difficult memories, the Paradise Theater may still stand today as a metaphor of the alternate worlds we temporarily choose to live in, in order to cope with the realities of the present."

Helguera could also have mentioned the beloved goldfish that swam in the Carrera-marble fountain near the candy counter. He could have described the Seth Thomas clock on the façade, where, as the hour chimed, a life-size equestrian statue of Saint George rose up on a miniature stage and slew a dragon that breathed fire, thanks to a flashing red light bulb in its mouth. And Helguera could have lingered indefinitely on the most ravishing special effect, the dark-blue ceiling in the auditorium that grew even darker before the show began, as if the sun were setting in a Cinemascope heaven, whereupon, one by one, twinkling stars appeared and filmy clouds drifted—pyrotechnics intended to give the nearly four thousand patrons sitting in the red velour seats at least the momentary illusion that they were watching a movie while luxuriating in a garden courtyard in the Mediterranean beneath a glorious moonlit sky.

Loew's Paradise—LOW-ees to generations of moviegoers—had not been imported stone by precious stone from an ancient Italian village. And although the architect whose vision gave life and shape to the theater was an immigrant, craftsmen from the New World, not the old, created its décor. Yet it was easy to understand why Helguera fell in love with the place. Loew's Paradise was a unique and magnificent creation, truly "an acre of seats in a garden of dreams," as one critic described the charms of such places. Others called it a "Wonder theater." Although "Wonder" referred not to the over-the-top décor but to the deluxe Robert Morton Wonder Organ that held pride of place in the orchestra pit, the phrase said a great deal about the allure of this particular dreamland.

$$\Longrightarrow \ \bullet \ \Longleftarrow$$

The Bronx's most famous movie theater arrived on the scene at an ideal moment. By the mid-1920s, some fifty million Americans were going to the motion pictures at least once a week, and the settings in which these films were shown—settings of "glamour, glitz, and sparkle," as film historian Jane Preddy calls some of the old picture palaces—were increasingly as much of a draw as the movies themselves. "We sell tickets to theaters, not movies," proclaimed the nation's venerable theater operator, Marcus Loew, in one of the most quoted pronouncements in the history of film, and in the spring of 1927, when the Fox Film Corporation opened the 5,920-seat Roxy near Times Square—"the cathedral of the motion picture"—the event ranked as one of the splashiest in the city that year.

The pictures shown in these theaters were a mixed bag; Hollywood always produced many more clunkers than people care to remember. Nevertheless, in a growing number of U.S. cities, even a dog of a film could make its debut in a jewel box of a setting. And although many talented and a few truly gifted individuals turned their hand to creating these theaters, John Eberson, the man who designed the Paradise, was in many respects in a class by himself.

From the start, Eberson led a colorful life. Born in Romania in 1875, he was educated in Dresden and at the University of Vienna. During those

John Eberson, the Romanian-born architect who created Loew's Paradise on the Grand Concourse just south of Fordham Road. The opulent movie theater, which opened in 1929, was intended to evoke the splendor of an Italian Baroque garden. (Theatre Historical Society of America)

years, recalled his son and longtime partner, Drew, "he had engaged in several duels, as was the custom in the fraternities of the European universities, and bore scars in testimony to what was considered friendly competition." Upon completing his studies, Eberson served for a year in the Fourteenth Hussars of the Austrian Army and was jailed after an altercation with an officer in his regiment. Escaping from prison, he fled to America, and by 1901 he had made his way to St. Louis, then home to a sizable German and Austrian community, to begin his life's work.

In a photograph that accompanies many of the articles about Eberson's long and prolific career, he resembles nothing so much as a stock character from some forgotten Viennese operetta. His square face is topped by a shock of silver hair and accented by a small mustache, elfin ears, and dark brows that shoot up at almost comical angles. A flowing four-in-hand tie peaks out of the collar of his crisp checkered shirt, the lapels of his jacket are cut uncommonly wide, a ring adorns one stubby finger, and his ever-present cigar is held at a jaunty angle.

In speculating on the inspiration for the lavish structures Eberson designed—more than three hundred theaters in the United States and abroad during a career that lasted until his death in 1954—his biographers invariably point to the environment in which he came of age, notably the rich folk traditions that flourished in the region of Romania where he was raised and the "fantasylike" culture of fin-de-siècle Vienna. Yet the man himself clearly had an innate taste for the unexpected flourish, the splashy detail. He delighted in the grand gesture. When he was living in Chicago and was told that owning racehorses was the mark of a sophisticated resident of that city, he promptly bought five at once; many years later, he arrived for a Christmas visit with his family accompanied by puppies for all the grandchildren. His writing, presumably intended to help potential clients share his own excitement about forthcoming projects, was as flowery as his designs, and just as irresistible. Along with creating showplaces, he imagined down to the most minute detail the lives that might have been lived within their walls had they been not playhouses but sumptuous castles and villas and mansions. In describing a new project in 1926, he wrote,

> I am working on a French interpretation of an atmospheric theater—
> the Garden of the Tuileries. We picture a Louis sending a message
> through the land calling to painters, sculptors, gardeners, artisans of
> all kinds. And he gives the command to transform the spacious lawns

lying in front of his palace into a festive ground, as he is going to entertain his grandees and dames at a glorious magic night feast. . . . Gigantic arches, enchanting colonnades, illuminated lattice garden houses, mystic pyrotechnic effects all silhouetted against the entrancing moonlit sky of a beautiful summer night. Surprises, illuminated fountains, music niches, lovers' lanes—a marvelous setting for a fantastic artful dance, the frills of the satin and silk-gowned nobles, the coquettish silk and ruffle-covered damsels, the air laden with the sweet perfume of jasmine.

Of his Avalon Theater in Chicago, a Persian palace for midwestern moviegoers, he wrote,

It is here that the royal nabobs and lords gather to barter and exchange everything from fruit to human souls. Behold the quietness and shade of this small pool where the numerous birds of brilliant plumage fly for refuge from the midday sun, to plunge unmolested in the clear cool water. Behold, the water carriers pause for a short siesta in the heat of the day to gaze in the pool of the Bridal Fountain and dream, perhaps—of more mysterious moonlight.

His daughter Elsa Eberson Kyle summed up her father's considerable talents with affectionate understatement: "He had a flair for the dramatic."

⇒ • ⇐

The seeds of Eberson's career were planted when he went to work for the Johnston Realty and Construction Company, a St. Louis firm that built opera houses in small towns around the country. In 1904, the man who came to be known as "Opera House John" started his own firm, and by the mid-teens he was designing the type of playhouses that made his reputation, among them a string of movie theaters in Texas and, in 1923, the Hoblitzelle Majestic in Houston, notable as the world's first "atmospheric"—a playhouse intended to make audiences feel as if they were watching a movie while seated in an exotic outdoor setting.

Eberson was hardly the only virtuoso of the picture palace during these years. His most talented contemporary, Thomas Lamb, also produced theatrical Xanadus, among them Loew's "Wonder theater" in upper Manhattan, praised for its "voluptuous pagan ornament" and "a rajah's ransom in furnishings," to mention only two of the attributes reporters singled out.

Nor was the atmospheric Eberson's only legacy; though he churned out dozens of theaters in that style—some estimates put the number at over a hundred—the list of his achievements was eclectic and included such structures as the band shell of Lewisohn Stadium in upper Manhattan and the Iraqi pavilion at the 1939 world's fair.

But the atmospheric, which was widely copied by other designers, quickly emerged as Eberson's signature creation, and all its characteristic innovations were on thrilling display at the Hoblitzelle Majestic. The auditorium evoked a courtyard in some unnamed Mediterranean country, possibly Greece, in that the décor included the Porch of the Maidens from the Erechtheum atop the Acropolis. Replacing the traditional domed ceiling was a blue plaster "sky" in which winking electric light bulbs simulated a starry heaven, and the impression of floating clouds was created by a machine in which film negatives were passed over light in a sequence of imaginative patterns. The device was called a Brenograph, described in an advertisement as equipped with a "universal electric motor with variable speed control with fleecy cloud effect complete in case . . . $290." Along with the mechanism that made the stars twinkle, the Brenograph lay at the heart of the theater's most alluring special effect, although not one moviegoer in a thousand had a clue as to how these devices worked.

Using such technical aids and increasingly lavish motifs, Eberson churned out one atmospheric after another. "We visualize a dream," he once wrote, "a magnificent amphitheater, an Italian garden, a Persian Court, a Spanish patio, or a mystic Egyptian temple yard, all canopied by a soft moon-lit sky." And shortly after setting up shop in New York City in 1926, Eberson began envisioning what became the grandest atmospheric of them all.

⇒ • ⇐

The original plan for the block-wide site at 2403 Grand Concourse called for a four-thousand-seat playhouse costing $1.8 million that was to be built by Paramount and called the Venetian, presumably in honor of a design scheme that featured a Venetian-Gothic–style façade. By the time construction began in early 1928, the original concept had been tweaked considerably. The Venetian was now the Paradise, the décor was now Italian Baroque, the budget had ballooned to four million dollars, and Paramount had turned the entire project over to Loew's, the chain headed by Marcus Loew that was building a group of theaters in the New York area featuring the Robert Morton Wonder Organ and was eager to include among their number a 3,884-seat playhouse that promised to be the largest, costliest, and most lavish picture palace in the Bronx.

The gala opening was set for the night of September 7, 1929, a moment that seemed propitious for many reasons. Another atmospheric, Eberson's Valencia, "a Spanish patio garden in gay regalia for a moonlit festival," as a publicity release described the place, had welcomed its first moviegoers earlier that year in Jamaica, Queens. Five weeks before the opening of the Paradise, on July 31, excavation began for Rockefeller Center, the city within a city poised to rise opposite Saint Patrick's Cathedral on Fifth Avenue. On August 4, Governor Franklin Roosevelt opened an oceanfront playground called Jones Beach on the southern shore of Long Island. And just four days before the opening of the theater in the Bronx, on September 3, Wall Street celebrated a record: the Dow Jones Industrial Average had soared to 381.17, then the highest point ever in the history of the U.S. stock market.

Whether these milestones were on the minds of the hordes milling outside the Paradise and spilling onto the boulevard that breezy, late-summer night was not recorded for posterity. The man from the *Bronx Home News,* the major paper represented, focused more on the extravagance of the décor than on the mood of the crowd. Nevertheless, given the sense of occasion, opening-night patrons were presumably in high spirits, dazzled by the visual fireworks served up by Eberson and company that were apparent even before they set foot inside the building.

The five-story façade of cream-colored terra-cotta, accented by touches of red Levanto marble, was a wild mélange of oversized windows, outsized columns, balustrades, arches, parapets, urns, cartouches, swags, Corinthian pilasters, and broken pediments, all of which made a fittingly ornate frame for the celebrated Seth Thomas clock. After taking a moment to observe the slaying of the dragon, the crowds politely pushed their way through the gleaming bronze doors, where they encountered a series of ornate spaces, each more sumptuous than the last.

The outer lobby, illuminated by a huge chandelier, was furnished with classical statuary interspersed with artificial shrubs on which perched feathery artificial birds—in reality stuffed pigeons. From there, patrons proceeded to the Grand Lobby, a great vaulted room patterned after the breathtakingly over-the-top Church of Santa Maria della Vittoria in Rome and furnished with plush chairs, alabaster lamps, and more chandeliers, the whole framed by walls upholstered in salmon-colored silk. Set in a golden ceiling and wreathed with cherubs and caryatids were three painted domes

The façade of Loew's Paradise, a mélange of architectural flourishes that was simply a warm-up for the extravagant décor within. (Theatre Historical Society of America)

depicting Sound, Story, and a deity unknown to the classical world: Film. The fountain in which goldfish swam depicted a child atop a dolphin and sat beneath a statue of the Winged Victory. On opening night, the gurgling of the fountain mingled with music from a baby grand piano.

At the foot of the palatial staircase leading to the mezzanine hung a series of works of art, among them an oil depicting *Marie Antoinette as Patron of the Arts* and a copy of Holbein's *Anne of Cleves*. Halfway up the

The inner lobby of Loew's Paradise, where an explosion of ornate details was surpassed only by the auditorium itself, a vast space where faux stars twinkled in a midnight-blue sky. (Theatre Historical Society of America)

stairs, a bronze statue of a young Hannibal kept guard, and on the mezzanine level, the by then slightly dazed patrons strolled along a promenade that stretched the length and breadth of the lobby on both sides and was crammed as far as the eye could see with statuary, urns, paintings, cherubs, and caryatids, the whole assemblage lighted by sconces of parchment-like hanging lamps splashed with Chinese calligraphy. The wooden phone booths were carved to resemble sixteenth-century sedan chairs; silk damask woven in France imparted a touch of the boudoir to the walls of the ladies' lounge, and lighting fixtures said to have been assembled from swords illuminated the men's smoking room.

All this opulence, however, was only a warm-up for the main event. Surrounding the vast vaulted space of the auditorium was every detail one could desire to suggest a Mediterranean courtyard: low garden walls, terraces, porches, balustrades, urns, more caryatids, niches, glowing palace windows—the whole adorned with artificial cypress trees, climbing vines, flowering shrubs, and a seemingly endless assortment of sculpture and statuary, including busts of Shakespeare and Wagner, statues of Venus, Apollo Belvedere, Cupid, and King Arthur, and a larger-than-life figure of Lorenzo de Medici. Above this extravagance floated the indescribably romantic sky, poised to darken and come alive with drifting clouds and pinpoint stars as showtime approached. "We credit the deep azure blue of the Mediterranean sky with a therapeutic value," Eberson once said, "soothing the nerves and calming perturbing thoughts." These "friendly stars," as Eberson called them, had been laid out by his son, who had joined his father's firm a few years earlier and who in arranging the stars followed a pattern "generally utilizing the January skies." In doing so, he depicted the heavens as they would have appeared at the time of the birth of Marcus Loew, who had died just two years before the completion of the Bronx picture palace that bore his name.

Such sumptuousness did not come cheap. A preliminary cost report prepared that December for the Concourse Realty Corporation, the company that built the theater, showed that in addition to $34,500 for the Wonder Organ and $6,750 for the clock, expenses included $9,443 for artificial flowers, trees, and shrubs; $25,000 for decorative furniture, much of which had been imported from Europe; and $168,000 for ornamental plaster. Yet despite the lavishness of the décor, the creation of the Paradise was in some respects an oddly personal affair, almost a domestic undertaking. Unable to find craftsmen who could paint the birds and the other ornamental details to his liking, Eberson created an entity called Michael Angelo Studios that in reality was his wife, an English-born decorator named Beatrice Lamb, who supervised a crew of skilled designers and craftsmen. With the help of the couple's daughters, Beatrice Eberson also assembled the heraldic banners that hung over the balconies while their son, Drew, tended to such mundane tasks as adjusting the pigeons and keeping an eye on the Brenograph, apparently a temperamental piece of equipment.

For the grand opening, the list of attractions, extensive even under normal circumstances, was unusually impressive. The singing of "The Star-Spangled Banner" was followed by music by the Paradise Grand Orchestra and a solo at the Wonder Organ. The stage show, Cameos, featured Dave Schooler and His Paradise Serenaders, the theater's resident singers, along with the Chester Hale Girls, a collection of leggy young women that included a vivid brunette named Kitty Carlisle who went on to marry a celebrated Bronx-born playwright named Moss Hart. Next came congratulatory messages transmitted from Hollywood by Loew's president, Nick

Schenk, and various stars associated with the Loew's-controlled MGM studio. Only well into the evening did the opening credits roll on the feature film, *The Mysterious Dr. Fu Manchu*, starring Warner Oland, Neil Hamilton, and Jean Arthur. Patrons in the balcony who were listening carefully heard the singing of live canaries, another festive touch to welcome the new playhouse to the Bronx.

≡ • ≡

To hear some people tell the story, no one in the West Bronx saw a movie anywhere but at the Paradise, and of course that was not true. The area was rich with theaters showing first- and second-run pictures, especially in the vicinity of bustling Fordham Road. And virtually every serious West Bronx moviegoer spent countless hours at the Ascot, a beloved art house that opened in 1935 at 183rd Street, a few blocks south of the Paradise. After sipping a tiny cup of inky espresso in the lower level—the Ascot also boasted one of the city's first coffee bars—moviegoers entered the long, narrow playhouse, there to get early glimpses of such classics as Marcel Pagnol's *Fanny* trilogy from the 1930s and Roberto Rossellini's *Open City* and *The Bicycle Thief,* two of the great Italian neorealist films from the 1940s, as soon as they arrived in the United States.

Programs at the Ascot changed only when business dwindled, and certain films were screened so often that the prints literally wore out. Jacques Tati's *Mon Oncle,* voted by the Academy of Motion Picture Arts and Sciences as the best foreign film of 1958, ran for nearly five months; the British New Wave film *Room at the Top,* the following year's winner, stayed for almost four. A passion for cinema defined the place. On Friday nights, the manager made an appearance to ask patrons if they'd liked what they'd just seen on the screen, and more than one moviegoer remembers the experience of literally being the only person in the auditorium.

"It was the theater of one's dreams," the writer Avery Corman recalled decades later during a stroll past the site where the Ascot once sat. "This is where you'd bring a classy date, to impress her and show that you were an intellectual." It was here that Corman saw Marcel Carné's 1945 classic *Children of Paradise,* David Niven in *Stairway to Heaven,* and, starting in the 1950s, the first films of Ingmar Bergman. "We weren't dumb kids," Corman said of the generation that cut its cinematic teeth at the Ascot. "We knew it was an intellectual mecca."

For Arthur Gelb, a former *New York Times* editor whose parents ran a children's dress store just a block away and who, like Corman, seemed able to recall the name of virtually every movie he saw at the theater, the Ascot as much as the Paradise provided an entryway to worlds unimagined in working-class and even middle-class Bronx. "We were always dreaming of living on Park Avenue," Gelb said. "The films we saw at the Ascot introduced us to the really sophisticated world that you only escaped to in the movies."

Nevertheless, the Paradise—given its size, its grandiosity, and its cachet—cast the largest shadow. You could not begin to count the number of girls who experienced their first serious kiss, or at least so remembered, in the sweaty reaches of the Paradise balcony. (That smoking was allowed in the balcony was a bonus.) But most of all the theater nourished a rich inner life and gave shape to aspirations. "You'd go with your grubby little dime," recalled Leonard Kriegel, who grew up near Mosholu Parkway and was another Paradise regular. "You'd get there at one o'clock and emerge at five or six bleary-eyed and totally strung out. The fantasy realism fed the imagination like nothing else I know of."

= • =

Even to work at the theater was a memorable experience. Gerald McQueen, born in 1937 in Virginia, the only child of deaf parents who never learned to speak, was among an endless series of young men and women employed as ushers at the Paradise. At the age of fifteen, McQueen ran away from home to live with a milkman uncle in the Bronx, and the following year he got a part-time job at the theater. Before long he was promoted to chief usher, a position that involved supervising a staff of dozens and brought a salary increase from fifty-five cents an hour to a dollar. A snapshot from those days shows a slim, dark-haired youth wearing a cardboard dickey and collar (changed after each shift), a jacket that seems too big for his lanky frame, and tan pants with a stripe running up the side. Serious and broad-shouldered, he crouches in front of an assortment of his charges and stares solemnly at the camera, the gold braid on his shoulders a testament to his lofty rank, his clip-on bow tie slightly crooked.

Working the four-to-midnight shift, McQueen stood smartly at attention as he admitted people on a line that often stretched around the block. Once the show started, he patrolled the darkened theater, flashlight in hand, keeping a particular eye out for members of the Fordham Baldies, who sometimes caused a commotion, and watched the elderly Italian man who stopped by occasionally to touch up the gilding.

= • =

Gerald McQueen, a Bronx boy who worked as an usher at Loew's Paradise in the 1950s. McQueen used to keep an eye out for the Fordham Baldies, a local gang, and watch as an elderly Italian man touched up the theater's famous gilding. (Private collection)

In countless powerful ways, the theater threaded itself into the texture and rhythms of the neighborhood. "To me, the Paradise was Sunday afternoons," the opera singer Roberta Peters, who grew up near the boulevard, once told a newspaper reporter. "We never were rich enough to live on the Concourse. We used to stroll on the Concourse—everybody did. And although I used to go to the neighborhood movies, the Luxor, the Kent, and the Zenith, Sunday was something else. First, we would all have Chinese food, and then we would go to the Paradise. What else was there to do on a Sunday?"

A regular visitor during those years would have noticed another habitué, a socially awkward young man with dark hair and piercing eyes, his chunky frame typically bundled up in a plaid wool mackinaw. He lived at 2715 Grand Concourse, and his name was Stanley Kubrick. Early on, he had developed a passion for black-and-white photography, and twice a week he showed up at the Paradise, as well as the nearby RKO Fordham, not simply to watch the double features but to analyze their style and structure. "I was taken with the impression that I could not do a film any worse than the ones I was seeing, and I also felt I could, in fact, do them a lot better," the filmmaker, by then a legend in his field, said several decades later. "One of the important things about seeing run-of-the-mill Hollywood films eight times a week was that many of them were so bad."

Bad or brilliant—for the vast majority of moviegoers, it hardly mattered. "You didn't check the papers to see what was playing; you just went," Corman remembered. After the movie, especially on Saturday night—date night—moviegoers automatically made their way to Krum's, the ice cream parlor across the boulevard from the Paradise, which at its peak had five soda fountains and seventeen soda clerks. Saturday-night crowds were so thick that people stood on line three and four abreast waiting for a place at the counter. "It's hard to believe it was not déclassé to take a date there," Corman recalled, "and that a sundae was a perfectly acceptable equivalent of dinner."

Those who didn't go to Krum's headed to Jahn's Old Fashioned Ice Cream Parlor, on nearby Kingsbridge Road. The cheerful, noisy ambiance featured fake Tiffany lamps over the wooden booths and such oversized offerings as "The Kitchen Sink" ("everything else but . . . serves four to six, $6.50"), "Suicide a la Mode," and "Sundaes, Mondaes, Schmundaes," names that made one's teeth ache almost as much as the hot fudge sauce. Another local favorite was Sutter's, the French pastry emporium at

Fordham Road and the Grand Concourse, whose elaborate Napoleons, éclairs, and cakes were visible behind cloudless panes of glass.

By general consensus, however, the best ice cream was to be had at Addie Vallins, an establishment located just west of the Grand Concourse at 161st Street and Gerard Avenue, and home of the famed black and tan, a coffee ice cream soda laced with chocolate sauce. Black and tans, the novelist Jerome Charyn writes in his memoir *Bronx Boy*, "were the trademark of Addie Vallins and couldn't be duplicated at another sugar bowl. A soda jerk would spend months on the magic formula: coffee ice cream that wasn't too creamy, the exact portion of seltzer and milk; and chocolate sauce with crushed pecans and a taste of bitterness that could have been the Bronx." Or perhaps it didn't happen exactly that way. Whatever the truth, even to this day Charyn insists that Addie Vallins did serve the world's best ice cream sodas.

"By the Waters of the Grand Concourse"

IN NOVEMBER 1945, in the wake of the Holocaust, the greatest cataclysm Jews the world over had ever known, a new magazine sponsored
by the American Jewish Committee burst onto the American scene. The
publication's name was *Commentary*, and although its official goal was to
engage young Jewish intellectuals in the great issues of the day, one of its
most beguiling features was a series of essays exploring what it was like to
be a normal American Jew living in relatively tranquil times.

The chatty and almost intimate pieces ranged widely in subject matter,
analyzing the cultural significance of everything from the neighborhood
pumpkin-seed peddler to that quintessential building block of the traditional Jewish brunch, the bagel. Several particularly vivid essays escorted
readers to the borough that had long since emerged as one of the great
centers of American Jewish life. "Bronx Housewife: The Life and Opinions of Mrs. Litofsky" charted the small dramas enacted in and near a
third-floor walkup apartment a few blocks east of the Grand Concourse,
and "The Real Molly Goldberg"—"the Baalebosteh of the Air Waves," as
the magazine called the patron saint of East Tremont Avenue—described
the version of this world with which much of America had fallen in love,
thanks to the television show that at its peak attracted up to forty million
viewers per episode.

A frequent contributor to this department was a young woman named
Ruth Glazer, a gifted writer whose lustrous good looks belied a gimlet-
eyed sensibility. Glazer had been raised in a series of apartments east of
the Grand Concourse, and half a century later, in a collection of essays on
the talismans of her old neighborhood, she confessed that she harbored
an affection verging on envy for the luxe of boulevard life.

Like countless Bronx children, Glazer was introduced to this world via
a medical mishap: when she broke her arm one winter, she paid several
visits to a doctor who maintained an office on the Grand Concourse. In
retrospect, these visits possessed an enchanted quality. "I would come

in the morning," Glazer recalled, "first passing through the hushed lobby and then taking the elevator to his apartment. Here I would sometimes see the son of the family—about my age—vanishing down a hallway in what seemed like illimitable space, having just picked up the newspaper that had been *delivered* to the door."

What a concept, she thought to herself—a newspaper gently deposited at one's doorstep on a snowy morning! This small act so captured Glazer's imagination that she endowed this doctor's family and their vast, hushed rooms with everything she longed for and found lacking in her own people's slapdash ways. In her eyes, the doctor's household reflected "the order, repose, and regularity of people who knew their way, who knew how to command the world as my parents did not, who knew, in fact, how to live."

Glazer (later Ruth Gay) is one of several essayists who sought to capture the texture of Jewish life on and near the Grand Concourse during its most intense period. For decades, the West Bronx was among the city's iconic Jewish neighborhoods, and especially in the years after the Second World War, the strengths and limitations of this world proved endlessly fascinating to writers.

The world was set in high relief by the fact that it did not exist in isolation. The neighborhoods of Highbridge, to the west of the Grand Concourse, and Fordham, to the east, were home to deep-rooted Irish Catholic communities, and a tiny cluster of Italian Catholics occupied the half-dozen blocks at the northwestern tip of the boulevard. Although African American families were largely barred from this solidly white part of the city until well after the war, black women were a local presence because so many of them were employed cleaning apartments on and near the Grand Concourse.

But it was Jews whose presence defined the West Bronx, and writers such as Ruth Glazer confessed to deep ambivalence toward their lifestyle and their values. Despite dreamy memories of her early visits to the doctor's office on the Grand Concourse, as a young and ambitious contributor to *Commentary*, itself a deeply serious and aggressively intellectual publication, Glazer painted a far more jaundiced portrait of the lush life on and near the boulevard. "Everywhere straight lines are abhorred," she wrote in "West Bronx: Food, Shelter, Clothing: The Abundant Life Just Off the Grand Concourse," an essay published in the magazine in June 1949. "Lamps are preferred in the form of baroque vases, their shades adorned with poufs and swaths of ruffling, the drapery is of a weight and quality calculated to set a luxurious barrier between the beholder and the University Avenue view. Wherever there is an upholstered surface, it is tufted; wherever a wooden one, it is carved into sinuous outlines and adorned with gilded leather." The crowning achievement, "the sign of a discriminating owner aware of the finer things in life," is the hand-painted oil that holds pride of place in the living room and to which there are only two appropriate reactions: the practical ("I'm telling you, the frame alone is worth it") and the aesthetic ("A picture like that, you won't get tired of looking at so fast").

Glazer was hardly alone in questioning the merits of this showiness; it was the rare West Bronx child who at least at times did not feel suffocated by rosewood and velvet and sickeningly plump comforters. And Glazer's assessment paled in comparison to that of Isa Kapp, another young contributor to the magazine, in whose eyes this gaudy excess represented not just a lapse of taste but a coarsening of values.

"At the threshold of the Bronx," Kapp wrote in an essay published later that year—a work that for decades was quoted to convey the ethos of this world—there emerged

> a Jewish community as dense, traditional, and possessive as William Faulkner's Yoknapatawpha County, and through it flows a great middle-class river, the Grand Concourse. There is no mistaking even its inlets and tributaries: the waters that seep over from the evergreened fountained courtyard of the Roosevelt Apartments to the modest tan brick of Morris Avenue carry an irrepressible *élan*, a flood of self-indulgence and bountiful vitality, vulgar and promiscuous, withal, luxuriant and pleasurable. . . . How many Concourse bedrooms are suffocated by flowered wallpaper, the *Kitsch* of domestic culture; how many windows blotted out with Venetian blinds, the somber instruments of urban privacy. The "living" room receives its sagging prop of barrel chairs, mahogany servers, cut crystals. In the spring, a housewife's fancy turns to chintz drapes with figures of birds and enormous roses.

"To confirm their faith in themselves and in America's promises," Kapp's description of this population continued, "they become conspicuous consumers of silver foxes, simultaneously of learning, gift-shop monstrosities, liberal causes, and Gargantuan pastries. A generous, expansive life! At the same time, a life utterly without taste. The rugs are too heavy, the spirit of the Jewish holiday is kept alive by fur pieces, the frame is always more expensive than the picture." Like her colleague, Kapp found few sins less forgivable than a fondness for bad art.

In retrospect, attacking the moral fiber of the entire West Bronx based on the local preference for flocked wallpaper and chintz seems a bit harsh. Nevertheless, in using the metaphor of a river to describe the boulevard— the essay is titled "By the Waters of the Grand Concourse"—Kapp, consciously or not, chose an image that brilliantly captured the notion of the street and its environs as a pulsing, quicksilver realm that defined and gave vitality to a great stretch of the city, one in which flood plains and waving grasses had been swept aside by shiny Deco and a grand roadway, yet nonetheless a place that exerted "a deep pull on our collective imagination," as a writer once described the power of the Mississippi. And the subtitle of Kapp's essay, "Where Judaism Is Free of Compulsion," perfectly nailed the essential nature of the most powerful tributary feeding this urban waterway.

═ • ═

For a part of the twentieth century, the West Bronx was so heavily Jewish that city officials determined how many Jews lived in the area using what was called the "Yom Kippur method," a system that involved counting the number of pupils absent from local public schools on the Jewish holiday. Yet the Judaism practiced along the Grand Concourse was very much a hybrid.

Although the rising and falling inflections of the Yiddish spoken by the grandparents echoed in the intonations of the next generation, although the rugelah, the seeded roll, and other comforting tastes of the ghetto had traveled intact from the Lower East Side and the East Bronx to the mahogany dining room tables of the Grand Concourse, and although marriage to a non-Jew still broke a mother's heart, much else had changed. The intimate little shul, where clusters of davening old men swathed in prayer shawls spoke directly to God, and presumably he to them, had long since been nudged aside, first by the more Americanized synagogue, then by the temple, and finally, in the postwar years, by the community center, a sociable place under whose capacious modernistic roof the chants of *baruch atah adonai*—"Blessed art Thou, O Lord our God"—were increasingly drowned out by the thump of the basketball and the chatter of the youth group.

Not all the traditions died, and not all at once. "Jewishness flickered to life on Friday night," Irving Howe wrote. "It came radiantly to life on Passover." Up and down the Grand Concourse and especially on the side streets, candles were lighted on Friday night, and there was no such thing as a nonkosher butcher. Nevertheless, one by one, and soon faster and faster, the ways of an earlier generation loosened their grip on this population.

If the grandparents fasted on Yom Kippur and spent the day at services, the grandchildren marked the holiest day in the Jewish calendar by showing up at temple at sundown, just in time to hear the blowing of the shofar and to walk their bubbe and zayde back to the apartment. Even holy men winked at the faith's dietary laws; a Grand Concourse boy who used to go on trips with friends who belonged to the Adath Israel youth group confessed three-quarters of a century later, "I never had the heart to tell Grandma that I saw the temple's rabbi, who usually accompanied the boys to the ball game, eat nonkosher hot dogs in Yankee Stadium."

The easygoing relationship these Jews had with the faith of their fathers had its roots in the German Jews who had settled the area in the early twentieth century and were far more assimilated than the second-generation Jews of Eastern European stock who followed after the First World War. Mostly, however, this attitude had to do with sheer numbers. The concentration of Jews in the West Bronx was so great that people who lived there had little need or reason to assert their faith. What else would they have been? they would have asked with a shrug as eloquent as any Sabbath observance.

Another factor was at work. As the historian Deborah Dash Moore, whose book *At Home in America: Second-Generation New York Jews* is the classic study of this population, explained in an interview,

> For these Jews, being Jewish isn't simply about God or about going to temple. Being Jewish is a spiritual, communal experience that involves families and especially intergenerational families—grandparents, aunts and uncles, lots of cousins. It involves one's whole being—the way you eat, the way you talk, your family, your neighborhood—all these secular aspects of life. What's Jewish for this group is less tangible because their Jewishness is inside them. But anyone coming to visit would know that they had come to a Jewish neighborhood.

═ • ═

In time, virtually all the congregations of the West Bronx migrated north to Westchester County or dissolved entirely, their houses of worship reborn as sanctuaries for Pentecostals or Baptists or Seventh-Day Adventists, the worn Hebrew lettering and the six-pointed stars on the façades the only faint reminders of their past lives. Yet in their day, and despite the gradual weakening of traditional observances, these institutions seemed both omnipresent and eternal. Even Jews who attended services rarely or not at all cherished their presence.

Morton Reichek, who moved to 1299 Grand Concourse in the late 1920s, when he was about four years old, remembers five synagogues on a single block of 169th Street, just steps from his apartment. On the northwest corner of 169th Street and Walton Avenue stood a synagogue that housed a Hasidic congregation. Farther up the block was his family's Orthodox shul, Tifereth Beth Jacob, whose rabbi, a white-bearded man named Levitan, was so holy-looking a figure that the young boy imagined that God in heaven looked like Rabbi Levitan. It was here that Mort delivered his bar mitzvah speech, in Yiddish, after which a dozen relatives gathered around the family's dining-room table to drink wine and nibble gefiltah fish. Next door to Tifereth Beth Jacob stood a Sephardic synagogue whose rabbi used to wander the neighborhood wearing a white turban and conducted services in Ladino, the language spoken by Spanish and Portuguese Jews. And at the southwest corner of the Grand Concourse and 169th Street, one of the premier intersections of the boulevard, stood Temple Adath Israel.

Adath Israel, which had started life in 1896 in a small private house in Morrisania, was said to be the borough's oldest synagogue. Notwithstanding an oddity in the basement, a portion of which had been leased to a tiny Orthodox synagogue composed of several dozen ancient men who prayed there mornings and evenings, Adath Israel was also the most celebrated Bronx synagogue, its fame due in part to the presence of a remarkable cantor.

He had been born Robin Ticker, or perhaps Ruvn or Rubin or Reuben Ticker—the biographical record is sketchy—and had grown up in a Williamsburg tenement, the sixth child of immigrant Jews. Thanks to a beautiful alto voice that ripened into a shimmering tenor, he was chosen by the cantor of a Lower East Side synagogue for serious study, and for a time, he worked as a part-time cantor in Passaic, New Jersey. In 1938, having reinvented himself as the more American-sounding Richard Tucker, he was hired by Adath Israel for an annual salary of twenty-five hundred dollars.

Five years later, the man who would reign for decades as the leading tenor of the Metropolitan Opera departed for a more prestigious synagogue in Brooklyn, largely, it was rumored, because Adath Israel's board refused to approve an annual raise of seven hundred dollars.

Even without a future star of the Met behind the bimah, this citadel of Conservative Judaism hardly lacked for panache. The white limestone neoclassical building, its façade flanked by Corinthian columns and pilasters, radiated a monumental grandeur that continued once worshippers passed through the great bronze doors. Just inside the entrance, such Old Testament images as the lion of Judah and the burning bush were pricked in brilliant stained glass. The vast, high-ceilinged sanctuary amplified the sound of the organ and the voices of the choir, whose members, dressed in academic robes, were invisible behind a screened balcony over the ark.

Not all local temples were this blessed. Young Israel of the Concourse, an Orthodox synagogue that started as a small shul on Walton Avenue, was so tiny that High Holy Day observances were held at the Concourse Plaza Hotel and was so financially needy that every year for two decades services included a fervent appeal for money for a new building, which was finally constructed on the Grand Concourse at 165th Street. Concourse Center of Israel, an Orthodox synagogue on the boulevard at 183rd Street, held its first High Holy Day services in a tent, although matters improved dramatically once real estate mogul Sam Minskoff was installed as head of the synagogue's building committee.

A small booklet published in March 1929, the souvenir journal marking the celebration of the Feast of Esther by the sisterhood of Tremont Temple, suggests the passions that fueled these communities. For generations, Tremont Temple ranked as the borough's leading Reform synagogue; the memorial plaques on the walls of its red brick Georgian building bore the names of turn-of-the-century German Jews who lived in private houses along the boulevard and arrived for services in horse-drawn carriages, the men wearing silk top hats.

By 1929, the congregation was less German, more Eastern European. Yet the effervescent publication marking the sisterhood's twentieth anniversary suggests that the synagogue's spiritual and social life had, if anything, grown more robust. The very names of sisterhood members, an array of Hetties, Birdies, and Roses, of Fannies, Lillies, and Sadies, conjure a generation of energetic, community-minded women. The rabbi's expansive praise of their contribution to the temple—"abundantly blessed through the devoted efforts of loyal Jewesses who have joyously serviced its cause"—expressed an almost tangible affection. And the page after page of advertisements, for Glickels at 1815 Grand Concourse ("Summer Furs and Foxes Our Specialty"), for Rosenhain's Pure Food Shop at 2469 Grand Concourse ("Caterer of the Exclusive Kind"), for Granite Lake and Queen Lake Camp ("for Jewish boys and girls"), and for 150 other businesses and organizations, offered vivid glimpses of how these women lived their lives, spent their money, and cared for their families.

≡ • ≡

Not every resident of the West Bronx patronized the local kosher butcher or lighted candles on Friday night. Some lived within the sound of church bells, religiously attended Sunday Mass, and sent their children to parochial schools where black-clad nuns enforced strict discipline long after neighboring public schools had relaxed their ways. Although Jews predominated on and near the Grand Concourse, the West Bronx was also home to several entrenched Roman Catholic communities whose religious similarities masked deep differences in ethnicity and economic status.

Preeminent among these communities were the Irish, who stood atop the city's political hierarchy and ran the great university on Fordham Road. Fordham had long been one of the city's most important Irish American strongholds, and Highbridge, the hilly area west of the Grand Concourse, had been settled by many Irish immigrants who labored on the aqueduct for which the neighborhood was named. Well after the apple and cherry orchards of Highbridge had given way to five-story walkups, the flood of Irish immigrants continued, so much so that at day's end, entire streets were perfumed with the smell of ham and cabbage and beef stew, the makings of a traditional Irish dinner.

So devout was this community that at times it felt like one enormous church. "I was never altogether certain of boundaries between the old country and the new or between this world and the next," the writer and professor Maureen Waters recalls in *Crossing Highbridge: A Memoir of Irish America*, her luminous chronicle of growing up on these streets in the 1940s.

> Our daily rhythms were punctuated by the beat of a celestial clock. Each morning we awoke to the tolling bell of the Carmelite convent across the street. It tolled the hour of Mass, the hour for prayer, ringing out for the Angelus at six o'clock, just as we were sitting down for

supper. It was a seductive, mournful sound, as mysterious as the nuns themselves, of whom we had rare glimpses as they moved about in fluttering white garments through their garden.

Having taken vows of poverty, the sisters were said to be hungry, and on holy days Waters brought them packages of food, in return for which they promised to say prayers for her. Sometimes she attended Mass in the convent's tiny gold and white chapel, entranced by "the soft rustling of garments behind a screen, the muted voices." In those moments, there seemed worse things a girl could do with her life than join this devout little world.

⇒ • ⇐

Although overshadowed by Belmont, the century-old Italian enclave centered on Arthur Avenue, near the Bronx Zoo, the Italian American community along Villa Avenue, a three-block stretch a block west of the Grand Concourse at the boulevard's northwestern tip, had its own claim to fame: poverty; a typhoid epidemic that broke out in its overcrowded shanties in 1904 was so severe that the news made the *New York Times*.

The community also remained close to its Old World roots, so much so that Masses at Saint Philip Neri, the local parish church, were recited in Italian until the Second World War. Religion was an immensely powerful local force; the highlights of the yearly calendar were the feast days honoring Saint Anthony in June and Saint Assunta in August, events that dated back to the turn of the century and were celebrated by uniformed marching bands parading along the Grand Concourse as banners for the saints flapped in the wind and the faithful pinned dollar bills to their statues.

And on a rainy November night in 1945, when an estimated thirty thousand people converged on a vacant lot just west of the boulevard, where a local nine-year-old named Joseph Vitolo Jr. claimed to have seen the Virgin Mary appear sixteen days earlier, the neighborhood seemed to stand at the very heart of American Catholicism.

The youngest son of immigrant parents, the boy had been born into punishing circumstances. His father, a garbage collector, was an alcoholic, and his mother had given birth to eighteen children, only eleven of whom survived past infancy. After supper on the evening of October 29, the boy had wandered from the family's apartment at 3194 Villa Avenue to a rocky ledge overlooking the Grand Concourse, and it was there, according to the breathless account published a few days later in the *Bronx Home News*, that he saw "the Virgin Mary with long blond hair and a sort of light around her." The apparition "wore a blue dress that turned to pink, and stood behind a golden table and four chairs," the boy added, and after handing him a candle, she promised to return for the next sixteen nights.

The father beat the son for telling lies. Nevertheless, it was too late to stop the flood of events that followed. As word of the apparition spread, fueled by newspaper articles and radio reports that rippled around the globe, crowds began descending, at first mostly members of the local parish but eventually from as far away as Cleveland. By early November, some three thousand people were arriving nightly, rosary beads clutched in their hands as they scooped up handfuls of dirt and proclaimed miracles of healing, many of which were promptly debunked. The Church hierarchy was generally dubious, as were some closer to home. Bernard Gwertzman, who lived around the corner, rolled his eyes at the memory. "We Jewish kids were very skeptical," Gwertzman recalled. "Very skeptical."

Notwithstanding naysayers, the frenzy continued, so much so that the child at its center was transformed overnight into an almost messianic figure. Instead of going to school, he held court in his family's living room, warmed only by a coal stove, where he attended the streams of visitors that included New York Archbishop Cardinal Francis Spellman and Frank Sinatra bearing a statue of the Virgin Mary. Catholics who had been sickened by incurable diseases or wounded by accidents and shrapnel were carried into the Vitolo apartment through an open window until shortly before seven in the evening, at which point the boy they had come to pay homage to was hoisted aloft on the shoulders of a relative and brought to the site of the apparition as people pulled at his hair and tugged at the buttons of his jacket. "I didn't understand what it was all about," Vitolo told a reporter more than half a century later. "People were charging at me, looking for help, looking for cures. I was young and confused."

By the final night, a sodden mess of rosaries, religious statues, trampled flowers, and hundreds of flickering candles carpeted the site. Women collapsed from hysteria, and special police blocked off traffic on a four-block strip of the Grand Concourse in a futile attempt to bring a little order to the spectacle that *Life* magazine had christened the "Bronx Miracle." Others used the term "mini-Lourdes," though reports were conflicting as to whether Joseph was familiar with *The Song of Bernadette,* the popular 1943 movie about another young bearer of glad tidings, a French shepherd girl who claimed to have seen a vision of the Virgin Mary in a local cave.

The events of those frenzied autumn days shadowed Vitolo's entire life. In 2002, by then a far less agile sixty-six-year-old, he was still living in his boyhood home and returning most evenings to the shrine that had grown up on the site overlooking the Grand Concourse where all the commotion had taken place. He still insisted that he had witnessed a miracle. "I never had any doubts," Vitolo, who at the time worked as a janitor at a nearby hospital and was the widowed father of two grown daughters, told a reporter. "Other people did, but I didn't. I know what I saw."

⇒ • ⇐

The conventional wisdom was that no African Americans lived on the Grand Concourse. That was not entirely accurate. Starting early in the twentieth century, black families could be found in what *Commentary* essayist Isa Kapp had described as a "miniature Negro slum," a cluster of tenements south of 149th Street, on the portion of the Grand Concourse that was originally known as Mott Avenue and never quite shed the stigma of being something of an afterthought. Many of these residents were southerners who had moved north in search of jobs and a more racially tolerant society, and at least one family had achieved a certain local renown: in the 1930s, a prominent local alderman named George Harris lived at 458 Grand Concourse along with his wife, Nettie, one of the city's first black female police officers.

But for the vast portion of the Grand Concourse's length, and during most of its existence, the boulevard was off limits to blacks in virtually every respect. This was an era of the de facto segregation that defined race relations in the North, a time when discrimination was blatant and pervasive in the borough and the city. Along the Grand Concourse, however, resistance to minorities was notorious.

Housing discrimination was so pervasive that when the Urban League sent teams of black and white investigators to test rental policies in buildings on the boulevard, the results were disheartening but predictable. Invariably, the black couple would be rejected immediately. When white couples were dispatched to the same apartment a week later, they would be offered a lease so quickly that they could have moved in the next day.

Although indignities came in all shapes and sizes, young black boys who seemed not to know their place were the most tempting targets. Jessie Davidson, a boy from the East Bronx whose family had moved to New York from Alabama and who in the late 1930s and early 1940s sometimes ventured over to the boulevard with friends, quickly learned that a police car would pull up almost the instant they set foot on the street. "The door would slam open and bang you right in your stomach," Davidson recalled in an interview conducted by the Bronx African-American History Project at Fordham University, and "the term nigger would come out of a policeman in a minute." Revolvers were drawn before any of the boys had uttered a word.

Virtually every black child who grew up in the neighborhood could tell a similar story. "You know, it's strange," Leroi Archible, a Memphis-born boy who carried his shoeshine box up the step streets that led to the boulevard to earn spending money, told an oral historian at Fordham. "Going west, and you went up those steps, which was Clay Avenue, you was in trouble. Once you hit the first step, by the time you got to the top one, you was in serious trouble."

To test the boundaries that defined this world was to tempt disaster. Around 1942, a black man named Edward Jones who worked as the superintendent in a building on Grant Avenue, two blocks east of the Grand Concourse, was found murdered, his naked body left lying on a steaming radiator in his apartment. The word on the street was that he had been killed in retribution for having had a relationship with a local white woman remembered simply as Miss Bobby. Jones's daughter, Dorothy, who had lived for a time in her father's building but moved with her two young children to a more congenial street in the East Bronx, discovered his body when she came back to pay a visit. It was never determined who killed him.

The starkest manifestations of the painful and complicated relationship between blacks and whites on those streets in those years was an outdoor job market that operated at various locations in the Bronx, including several near the Grand Concourse. There, black women seeking jobs cleaning apartments waited to be hired by women who were white and who were almost all Jewish.

The market was born during the Great Depression, a period that had uneven impact around the city. For the Jews of the Grand Concourse, even the worst years of economic collapse did not bring automatic hardship. This is not to say that the West Bronx escaped unscathed. Even formerly well-situated families bounced about during those years, moving from one apartment to a slightly cheaper place to cash in on offers of a free month's rent. While families remained largely intact, many adult sons and daughters were trapped in their childhood bedrooms long after college and even after marriage, an arrangement that at best proved confining and at worst claustrophobic.

The institution known as the Bronx Slave Market, a street-corner job mart that operated at various locations not far from the Grand Concourse, at which African American women waited to be hired by the white, mostly Jewish matrons of the West Bronx. The institution, immortalized in a series of photographs taken in 1937 by Robert H. McNeill, endured into the 1950s and became a symbol of the racism and inequality that ultimately exploded onto the boulevard itself. (Estate of Robert H. McNeill)

But the image of the evicted family huddled on the street, its pathetic belongings strewn on the sidewalk, was not part of the Depression-era iconography in this part of town. Although hardly as wealthy as Park Avenue or even Central Park West, this was a prosperous neighborhood, especially the portions on and near the boulevard, and Black Tuesday and its aftermath did not entirely or permanently erase that prosperity. Even among less well-to-do households, the fact that so many Jewish wives and mothers were comfortable working outside the home did much to cushion the impact if a husband or father lost a job. Many families, and not just the truly wealthy, kept their heads above water economically. They hired housekeepers, they took vacations, and their offspring remained in school, especially at the city's extensive network of free colleges. Although

The Bronx Slave Market was christened by Marvel Cooke, an investigative reporter who went undercover to report on its activities. The photograph's ironic title, "Make a Wish," is taken from the sign in the background. (Estate of Robert H. McNeill)

residential construction along the boulevard ground to a halt after the market crashed, development roared back long before the Second World War brought an official end to hard times.

For the great majority of the city's black families, who were among those disproportionately affected by the city's soaring unemployment rate, and especially for the black women who offered their services at the street market, the picture was far grimmer.

The market, christened the "Bronx Slave Market" by the black journalist who brought it to public attention, was one of an estimated two hundred outdoor job marts around the city and was a response to the role of the *schvartze*—"the black one," in Yiddish—the nameless, faceless black maid who was a fixture in comfortable Jewish neighborhoods such as the West Bronx. Sometimes housewives recruited her informally; the invitation "Girlie, you looking for a day's work?" could be heard drifting from apartment windows that faced the Grand Concourse when a woman with dark skin passed. More often, however, her services were secured at street markets such as those at the corner of Simpson Street and Westchester Avenue, about a mile east of the boulevard, and at several just a few blocks west of the boulevard—at Jerome Avenue and 167th Street, at Gerard Avenue and 167th Street, and at Walton Avenue and 170th Street.

Sometimes in the rain, often in the cold, women who came mostly from Harlem and started their trek around dawn waited at these intersections to be tapped by a Bronx housewife for a long day of grueling labor. Rates varied: seventy-five cents an hour, twenty-five cents, sometimes as little as ten cents an hour—a pittance even at a moment that the average New York family earned well under two thousand dollars a year. Frequently, the lady of the house set the clock back so as to trim these meager wages even further.

The Bronx market is little remembered. Nearly all the women who took part have died, and because offering one's services in this fashion was a source of humiliation, they rarely spoke to their children about this chapter in their lives. Yet though the market died more than half a century ago, the practice represents a crucial strand in the history of the Bronx and especially the West Bronx. By its very existence, the presence of the market underscored how separate were the lives led by whites and blacks in those years, a separation born not just of race but also of income and class. In many respects, the market represented the deep secret of the Grand Concourse, embodying a hidden poverty and inequality that remained unaddressed and that in the years to come would dramatically fuel the area's decline.

Two eloquent documents trace the lineaments of this cheerless world. One is a series of photographs taken in the late 1930s by Robert H. McNeill, a prominent photographer of midcentury African American life. The day McNeill took his camera to the Bronx must have been a cold one; his subjects are swaddled in heavy winter coats, their heads emerging from large fur collars. Several are wearing hats topped with stylish little feathers, a nod to the fact that these women sought to make a good first impression, despite the fact that prospective employers looked mainly for calloused knees as testimony that a woman had spent many hours scrubbing floors. The women stand and visit among themselves or sit on upended vegetable crates. Their expressions, what little we see of them, are inscrutable.

The other document is an article by Marvel Cooke, the intrepid journalist who went undercover to report her story, which was published in 1935 in the magazine the *Crisis*. So powerfully did Cooke describe the scenes she witnessed that many decades after the article first appeared, badly Xeroxed copies of her piece were passed from hand to hand like samizdat among people with an interest in black history, a reminder of a grim moment in the city's labor past.

The author of this remarkable piece of reportage, who died in 2000, just shy of her one hundredth birthday, was a gutsy, outspoken woman whose deep-set dark eyes and carefully coiffed hair gave her a glamour that belied some of the environments she found herself exploring in her work. Born in Minnesota to a politically active family—her father was a Eugene Debs Socialist—Cooke moved to New York in 1926 shortly after college and began her career as a secretary to W.E.B. Du Bois, who sixteen years earlier had founded the *Crisis*, a publication of the National Association for the Advancement of Colored People. Cooke later worked as a reporter for the *Amsterdam News*, at which time she joined the Communist Party and was jailed for helping organize a chapter of the Newspaper Guild, and for the *Daily Compass*, where she was the only woman and the only black person on the staff of the publication that was the successor to the left-wing newspaper *PM*.

In an era in which even white women reporters were a rarity, Cooke wrote about a wide variety of subjects, from prostitution to the Rosenberg espionage case. Yet so haunted was she by the conditions she witnessed at the street mart she christened the "Bronx Slave Market" that she returned to the subject again and again, describing it for various publications over the years. "I will never forget that experience as long as I live," she wrote after one visit. It was, however, her first article on the market, written with a black activist named Ella Baker, that most powerfully stirred the minds and hearts of her readers:

> We invaded the "market" early on the morning of September 14. Disreputable bags under arm and conscientiously forlorn, we trailed the work entourage on the West side "slave market," disembarking with it at Simpson and Westchester avenues. Taking up our stand outside the corner flower shop whose show window offered gardenias, roses and the season's first chrysanthemums at moderate prices, we waited patiently to be "bought."

Along with offering her own services at this corner, she talked to young women such as Millie Jones, who described her experiences working for a West Bronx housewife named Mrs. Eisenstein. Millie Jones told of scrubbing and then waxing floors on her hands and knees, washing fifteen windows inside and out, beating mattresses, and every week laundering and ironing the twenty-one shirts accumulated by Mr. Eisenstein and the family's two grown sons. Cooke herself was ordered to scrub floors using only a washcloth doused in a bucket of water and was paid seventy-five cents an hour for her efforts. For her, too, the clock was set back an hour, so as to pare her wages even further.

Officially, the Bronx Slave Market ended in 1941, when Mayor LaGuardia established the Simpson Street Day Work Office in the East Bronx, a domestic employment bureau where hours and wages were regulated and services were provided without charge. The market reappeared after the war, however, and as late as 1950, Cooke was going undercover yet again to report on the problem. Only in the late 1950s did the system finally wither, a victim of an improved economy and the advent of the civil rights movement.

7

The Grand Concourse of the Imagination

MANY MEMOIRISTS and a few novelists have drawn compelling portraits of what it was like to live on and near the Grand Concourse during the years of its greatest fame, portraits ranging in tone from grim to sugar-coated. One of the most revealing depictions, however, takes the form of a long-out-of-print first novel that hardly anyone remembers and even fewer people have actually read.

The work, set in the final years of the Depression and published in 1954, is called, simply, *Grand Concourse*. Without the Internet, one might stumble upon the novel only accidentally, perhaps in a secondhand bookstore or in the stacks of a library that never throws anything away. Even its author, an aging widower named Eliot Wagner who lives in upstate New York, in a bucolic setting that bears little resemblance to the Bronx streets of his youth, acknowledged in an interview a few years ago that *Grand Concourse* made barely a ripple on the American literary scene. "There were so few copies," Wagner said in a conversation about the ups and downs of the writing life. "Maybe five thousand. It died quickly."

Tepid reviews didn't help. Critics found *Grand Concourse* lacking in literary merit, the setting dreary, and the characters "doomed in advance," as the headline over the review in the *New York Times* summed up their unhappy fate. Critics also faulted the author for being more stenographer than artist, unsurprisingly, given the source of the book's inspiration. Wagner freely admitted that he drew all the characters from real life, and he himself lived so familiar a Bronx existence that he could have made a cameo appearance in his own narrative.

The son of a mailman, Wagner was born in 1917 on Fox Street in the East Bronx and grew up a few blocks off the Grand Concourse. After De-Witt Clinton High School, he attended City College, where he studied creative writing with Theodore Goodman, one of the institution's legendary professors. But Goodman did not encourage this particular student, and in any event, writing did not represent an easy career choice for

Depression-era graduates. After three years of unemployment, and with jobs hard to come by, Wagner went to work as an administrator for the city Board of Transportation, where he stayed until his retirement, writing novels on the side.

Despite the book's obscurity, *Grand Concourse* is an unexpectedly moving work, peopled by characters whose lives are measured almost entirely by their proximity to or distance from the thoroughfare of the title. The lovely and sensitive Julie, daughter of a struggling grocer, lives on Tiffany Street in the East Bronx, as does her brother Gerald, the aspiring writer who works as an usher at the Excelsior, the novel's stand-in for Loew's Paradise. Sam, the unhappily married civil servant, lives around the corner on Fox Street. Clara, the girl with whom Gerald is infatuated, lives on the Grand Concourse proper, and Sam's sister, Deborah, who married well, has made her way to a mansion in the gilded green precincts of Riverdale. The vast gulf between the East Bronx and the West is bridged mostly in dreams.

Not withstanding the novel's title, the boulevard is an elusive and almost mysterious presence, "vast" and "shadowy," as Wagner puts it, there and not there. The street makes only fleeting appearances, mostly when characters cross it to travel from one realm to another, yet it stands as a constant and powerful reminder of the gap between the lives lived on the Grand Concourse and those lived off it.

> The Grand Concourse, the broad boulevard crowning the center of the three ridges of the Bronx, its West End, its Park, its Fifth Avenue, meant to Gerald only the changing to an uptown bus. . . . [He] dropped into a seat and looked once more at his watch. God, how late!
>
> Had he been able to call on Clara at home, the lateness would not so much have mattered. But it was the street corners for them, on account of her mother. . . . [He] watched the passing line of apartment houses, some in red brick, some in tan, some few in expensive gleaming white—but all respectable! Oh, yes, respectable. Clara's mother, who lived in one of the most modest Grand Concourse dwellings, had once gone so far as to wave her lease in Gerald's face and shout, "What are you doing here bothering my daughter? Get back to Tiffany Street where you come from!"

⇒ • ⇐

In teasing out the role the Grand Concourse played in the literary imagination, it would be simplistic merely to tick off those writers who have mentioned the street in their work. In some recent crime fiction, for example, the Grand Concourse serves only as a colorful and instantly recognizable backdrop intended to signal the presence of the mean streets against which the drama will unfold. But the work of other writers, even some who relegate the boulevard to just a cameo role, proves unexpectedly revealing and helps bring into focus the various and complex ways the street functioned as a public and often iconic place.

One name invariably cited when people speak of fiction set in this part of the city is Arthur Kober, with particular reference to his sketch "The Daring Young Man on a Bus," first published in the *New Yorker* and later collected in *Thunder over the Bronx*. While Wagner's mostly tragic rendering of life lived on the fringes of the Grand Concourse disappeared from public consciousness in a heartbeat, Kober was far more fortunate in his legacy. If ever a single image of the Bronx lingered in the minds of mid-century Americans, at least those sophisticated enough to subscribe to the *New Yorker*, it was the Bronx inhabited by the irrepressible Gross family and particularly their unforgettable twenty-one-year-old daughter, Bella, all of whose lives revolved around the pursuit of "a fine steady boy who knows how to put by a dolleh." Or as Mrs. Gross puts it when dismissing one of her daughter's "platonic" friends, "Tonic-schmonic. Believe me, all I say is when I see my dutter married, I'll be happy like anything."

Kober, the creator of what his *New Yorker* colleague Dorothy Parker described as "dolling people," was a small, serious, and extraordinarily shy writer whose ability to produce the baroque idiom typical of certain Bronx neighborhoods was honed through a series of unlikely circumstances. Born in 1900, Kober came of age in the borough, but his schooling ended after one semester at the High School of Commerce, and he later contended that his inimitable style sprung from a lack of formal education that made writing a single declarative sentence a nightmare. After stints as a theatrical press agent, he went to work for an author of inspirational books bearing such titles as *A Thought for a Day*, not long after which Kober was talking uncannily like Bella Gross. "I never read a book, I'd peruse it," he once told an interviewer. "And I never agreed, I'd concur. And I never said yes, I acquiesced."

Kober, who ended up in Hollywood writing screenplays and the hit Broadway play *Having Wonderful Time* and who married the playwright Lillian Hellman, never specifies exactly where in the Bronx his sketches take place. (The composer Harold Rome, who wrote the score for *Wish You Were Here*, the musical version of Kober's play, was far more specific when he composed the lyrics for Jerome Weidman's musical *I Can Get It for You Wholesale*, especially in the song that begins "When Gemini meets Capricorn / Right across from the Grand Concourse / Could be an accident / Could be some heavenly force."). Still, the Grosses know which landmarks signify social success. When Bella latches onto "a brannew boy," her mother predicts that "soon will be by us a big wedding on the Concuss, mit a foist-class son-in-lawn." When Mrs. Gross is redecorating the family's apartment, her landlord advises, "Go to the finest flets on the Concuss and is oney one color—chotruse." (The landlord has already wisely warned her off crimson: "You know where you find crimm walls? In the chipp apottments where is living very common pipple.")

And it is aboard a double-decker bus heading up the Grand Concourse that Bella meets one of her more presentable conquests:

> She was unaware of the snappily dressed young man with the black, toothbrush mustache when he sat down beside her. She therefore gave a violent start when he addressed her.
>
> "I begya podden, Miss," he said, pressing his hat brim, "but haven't we met before?" He smiled so broadly she could see his gold fillings. . . .
>
> Bella would ordinarily have risen and walked away. Something about this young man's manner, however, something about his clothes, indicated that he wasn't what she called "the type boy who gets fresh with a girl and right away he commences making field day with his hands."

Almost from the moment Bella first sashayed into the living room of her family's modest apartment, wearing a sweater that she hoped didn't show "too much—y'know—the bust, excuse the expression," Kober had his detractors. Many critics took aim at his little love songs to the borough and particularly squirmed at his dialect, thick as a serving of Mrs. Gross's legendary "spuntch cake." He was dismissed as a vaudevillian and his concoctions as burlesque. Isaac Rosenfeld, the golden boy of the New York literary intelligentsia but a critic playful enough to have written a Yiddish spoof of T.S. Eliot's *The Love Song of J. Alfred Prufrock*—"ikh ver alt, un mayn pupik vert mir kalf" (I grow old, and my bellybutton grows cold)— admitted in *Commentary* magazine that he loathed the Grosses and everything they represented.

Others of equal stature, charmed by Kober's humor and his ear for a certain sort of Bronx speech, disagreed. In a loving obituary published in 1975 in the *New York Times*, the writer Israel Shenker, no slouch as a stylist himself, concluded that Kober "was to the Bronx what Erskine Caldwell was to Georgia, William Faulkner was to Yoknapatawpha County in Mississippi, and James Joyce to the unconscious." Kober himself confessed a deep affection for his characters, perhaps, he said, because they reminded him a little of himself, particularly the untutored young man he had been when he was Bella's age. "I consider the people poignant," he told an interviewer. "This surface elegance . . . covers up a deficiency, and it is sad and wistful. At one time I had a superior attitude toward them. Now I don't."

= • ⇒

When Eliot Wagner and Arthur Kober were writing, the world that was their subject matter was changing in only incremental fashion. Even as the city sloughed off the remains of the Depression and struggled through a world war, the basic tenor of life in the West Bronx remained steady and if anything more robust.

That was not the case a generation later. When writers born in the 1930s looked back to the Bronx of their early years, they gazed on a world that had all but evaporated. In this, they had much in common with American Jewish writers born in the teens, among them Saul Bellow, Bernard Malamud, and Delmore Schwartz, whose subject was frequently the immigrant world. The critic Irving Howe compares this earlier generation to its counterparts in the South, writers who also turned their gaze on the world they knew best. "In both instances," Howe concludes,

> a subculture finds its voice and its passion at exactly the moment that it approaches disintegration. This is a moment of high self-consciousness, and to its writers it offers a number of advantages. It offers them an inescapable subject. . . . It offers the emotional strength which comes from traditional styles of conduct. . . . It offers the lure of nostalgia, a recapture of moments felt to be greater and more heroic than the present. . . . [It] offers the rhythm of exhaustion, a way of life coming to its end.

The intersection of what Howe calls "receding cultures and obsessed writers" recurs toward the end of the twentieth century, a period that also produced a generation of creative artists whose subject was a society close to extinct. Theirs too was fiction as elegy, and although many immigrant neighborhoods provided grist for reverie, the world of the Grand Concourse held particular allure.

Its most powerful and glowing rendition can be found in E.L. Doctorow's 1985 novel *World's Fair*, the tale of a young boy who experiences momentary glory in a visit to the 1939 extravaganza in Flushing Meadows Park. The story has the style and shape of a memoir; it is told in the first person and laced with autobiographical detail. Like the author, the hero is a boy named Edgar, born in the West Bronx in 1931; his parents are Dave and Rose, he has a brother named Donald, and he lives on Eastburn Avenue, a few blocks east of the Grand Concourse.

Because Doctorow writes with such familiarity and precision about the world in which he grew up, he is often asked how much he has drawn from his own life, and invariably he responds the same way. "The book is an invention," he says. "It's the illusion of a memoir." Nevertheless, in *World's Fair* he describes an environment he knew intimately, and some of the novel's great pleasures are the closely observed descriptions of the texture of life on the Grand Concourse in the years before the war and the deep emotional pull this life exerted on the novel's young narrator.

> My grandma and grandpa on my father's side were not people of means, living in a three-room apartment a few miles to the north of us. But they were whole, complete, they took pride in themselves and their children. . . .
>
> We rode up there on the red-and-black Concourse bus, which had a long engine hood and doubled rear wheels and spare tires chained to the back and torturously shifted gears. . . .
>
> My grandma liked to put out a lace tablecloth on the big dark table in their living-dining room. She served tea in her good china with its pale green and white sliced-apple motif, and also Uneeda crackers and homemade plum jam with cloves in it and a big cut-glass bowl of fruit, and a smaller bowl of pistachio nuts. Most often too we brought cake from a bakery. And everyone sat around the table and talked. My grandfather had a wonderful way of paring an apple, with his own pocketknife, so that the peel came off in one continuous strip. . . .
>
> Another diversion was the dumbwaiter in the kitchen. My grandmother let me open the little door in the wall and poke my head into the black air shaft. Odors of ash and garbage rose on the cold black air.

> A thick rope bisected the column of darkness. I could pull on this rope and bring into view the wooden box on which the tenants delivered their garbage to the superintendent.

The narrator's grandparents are established people, his grandmother with her tight yellow-white bun, his grandfather with his tan cardigan, a retired printer from Russia, a Socialist who told his grandson that he had voted for William Jennings Bryan in three presidential elections.

World's Fair demarcates a clear line between life on the Grand Concourse and off. Nevertheless, Doctorow's own situation helped him understand that "while it was a reflection of immigrant aspiration, of dignity and class, to live on a thoroughfare as grand as the Champs Elysees," that was not the full story. In the early 1940s, when Doctorow's father fell on hard times, the family moved from spacious quarters in a private house on Eastburn Avenue to a three-room apartment on the boulevard at 175th Street. "It was clearly a step down," Doctorow said. "We felt it was a step down."

Doctorow suggests the limitations of this life in "The Writer in the Family," a short story in which he returns to the characters who inhabit *World's Fair*. Although the story appeared a year earlier than the novel, its characters have aged, and badly. The father, Dave, an appliance salesman, has just died, and his widow is living with her two sons in a cramped, three-room apartment on the Grand Concourse, a warren stuffed with bad memories, ill feelings, and the stench and detritus of recent death.

> I shared the bedroom with my brother. It was jammed with furniture because when my father had required a hospital bed in the last weeks of his illness we had moved some of the living room pieces into the bedroom and made over the living room for him. We had to navigate bookcases, beds, a gateleg table, bureaus, a record player and radio console, stacks of 78 albums, my brother's trombone and music stand, and so on. My mother continued to sleep on the convertible sofa in the living room that had been their bed before his illness. . . . There were lots of appliances in the kitchen—broiler, toaster, pressure cooker, counter-top dishwasher, blender—that my father had gotten through his job at cost. A treasured phrase in our house: "at cost." But most of these fixtures went unused because my mother did not care for them. Chromium devices with timers or gauges that required the reading of elaborate instructions were not for her. They were in part

responsible for the awful clutter of our lives and now she wanted to
get rid of them. "We're being buried," she said. "Who needs them?"

⥸ • ⥷

For the most part, Doctorow re-creates the streets of his childhood with
enormous affection. But if any single piece of writing could be considered
a valentine to the Grand Concourse, it is Avery Corman's 1980 novel *The
Old Neighborhood,* whose heart is a lyrical evocation of the author's West
Bronx childhood. Like Corman, the protagonist, Steven Robbins, came
of age during the 1930s and 1940s in a walkup apartment just off the
boulevard, and among his seemingly endless memories are those of the
rituals enacted along the street's broad expanse, none more potent than
the bursts of patriotism that followed the Second World War. Although
Robbins, like Corman himself, was too young to experience directly
much of the fear and heartache of the war years and, like so many West
Bronx children of his generation, was shielded from the conflict's greatest
hardships, he was well placed to savor the euphoria that accompanied the
peace.

> The Memorial Day Parade along the Grand Concourse after V-E Day
> was a massive victory celebration. Everyone who could march was
> there—servicemen on leave wearing their uniforms, civilians from
> war organizations, the wounded in special cars. Those families whose
> windows faced the Concourse competed with each other in the size of
> their American flags. Arthur, whose apartment was on the Concourse,
> invited us to watch from his living-room window. Whenever a color
> guard passed by, the spectators sitting along the curb would rise and
> come to attention. Uncertain of protocol and wanting to do the cor-
> rect thing on this important day, when a color guard passed beneath
> us, we stood at attention in the living room. . . .
>
> I was seven years old when the war began and eleven when it
> ended. Beyond the imagery, the war movies, the war posters, the war
> fantasies, what I remember is the sense of community in the neigh-
> borhood in those years. It was special and profound and I have never
> forgotten it.

Robbins travels a predictable journey. After graduating from DeWitt
Clinton High School and studying business at City College, he lands a job
in advertising and is soon possessed of a flourishing career and a picture-

perfect family. After a time, however, his seemingly idyllic life begins to unravel, and his thoughts return to the streets of his childhood.

> I replayed stickball games. . . . Memory taunted me. I wanted all of an experience that was good, an entire walk through the neighborhood with Arthur and Jerry, every word said, every nuance, an entire Saturday night date in the Bronx, everything that happened, every part of the physical sensations—the hot fudge sundae at Krum's Soda Parlor, the warmth of Cynthia Cohen's crotch as I forced my hand underneath the crinoline beneath her skirt and tried to work my fingers below her moist panties.

Shakily, Robbins makes his way back to this world, where he regains his equilibrium by playing half-court basketball at the local playground, whipping up malteds at the local candy store, now fallen on hard times, and trading wisecracks with the neighborhood bookie, who always regretted that a child with so remarkable a head for sports trivia never followed in his professional footsteps. Yet even as Robbins is briefly anesthetized by familiar terrain, he realizes soon enough that the landscape of his childhood has been virtually obliterated and that even many recent arrivals on the scene yearn to move on. By the novel's end, he has found solace in a simulacrum of this memory-glazed world, an antiques shop called the Old Neighborhood, where he summons ghosts of the past by trading in model trains, Depression glass, and other talismans of his childhood.

In offering a lyrical evocation of a certain sort of West Bronx childhood, the novel provides an almost preternaturally sharp picture of a particular community at a particular moment. As vividly and apparently artlessly as a snapshot, *The Old Neighborhood* captures the years during and after the war when in retrospect, and sometimes in reality, many parts of the city offered at least some of their residents the sense of community and connection to a place that is really what people are talking about when they wax nostalgic about all those egg creams and stickball games and long-ago baseball players.

In part because *The Old Neighborhood* drew so generously on the texture of Corman's early life, the author has emerged as something of a go-to guy when it comes to Bronx nostalgia—a Bronxologist, as he described himself in an article in the *New York Times*. He has also conducted tours of the neighborhood, stopping at such landmarks of his youth as the Jacob H. Schiff Center on Valentine Avenue, where he and his neighbors prayed for the safe return of Allied soldiers on D-Day, and the Love Gospel Assembly, formerly the Concourse Center of Israel, where his bar mitzvah was celebrated. Much like his protagonist Steven Robbins, Corman once went so far as to whip up egg creams in a candy store near his old apartment, an establishment by then run by Koreans.

On these strolls down West Bronx memory lane, Corman serves as so genial a narrator and is so visibly moved by a return to familiar streets that his would seem to have been an idyllic childhood; unspooling in such a storied setting, his audience might assume, how could it have been otherwise? That was not entirely the case. Corman's parents were divorced when he was young, and he grew up in a single-parent family, a rarity in those years. The household also included a rotating assortment of cousins and a deaf aunt and uncle, not the easiest environment in which to come of age.

Corman, like Doctorow, learned firsthand that even during the boulevard's happiest moments, and despite its lustrous reputation, the Grand Concourse never represented an unbroken strip of elegant buildings occupied by happy and prosperous families; the reality was always more complicated. Interspersed with those handsome Art Deco apartment houses and in many stretches outnumbering them stood large numbers of five-story and six-story walkups, home to families of far more modest means. In those years, few things signaled one's social class more than the presence of an elevator, and every time residents of those buildings trudged up and down their narrow and dingy stairwells, carrying babies and groceries and bundles, they were reminded in wearying fashion exactly where they stood in the pecking order of the West Bronx.

Corman came to understand these logarithms of neighborhood life in a profound way in 1953, the year his family moved from a five-story redbrick walkup on Field Place, a tiny street that jutted off the boulevard north of 183rd Street, to a modern apartment house with an elevator at 1695 Grand Concourse, near Mount Eden Avenue. "That was my old apartment, the third window," Corman said one day a few years ago as he stood on Field Place and pointed up at his building. "The windows on the side entrance overlooked the Concourse, but I would never say to people that I lived on the Concourse. To me, that meant something else entirely."

Like the area near Southern Boulevard where Corman lived before his family moved to the West Bronx, the streets around Field Place were home not to those who had already established a firm toehold in the middle class but to strivers. "My building looked more like the tenements of the East Bronx, where we used to live," Corman said, "and to be honest, I felt more in common with that world than with the doctors' sons. We all went to the same schools, but you could live on the Concourse and not be a rich kid. We knew there were economic differences. We knew where the rich kids lived."

The move to 1695 Grand Concourse brought a keen sense of having journeyed to a very different part of the boulevard, despite the fact that Corman's family had traveled little more than a mile. For the first time in his life, Corman lived with such amenities as a lobby, new kitchen appliances, and a bathroom without a claw-foot tub. For the first time, his mother felt comfortable inviting friends over to visit. "The move to 1695 Grand Concourse represented my mother's attempt to improve her station in life," Corman said on that memory-drenched day in the Bronx. "But looking back, it's astonishing to me that this was considered a cultural step up."

After college, Corman worked for a time in the advertising business, and as he began to make his way in the world, he discovered another unspoken truth about West Bronx life—that even the boulevard's most privileged residents lived in a setting far removed from Manhattan-style luxury and that even a berth on the Grand Concourse took one only so far. Even the Grand Concourse did not represent the ultimate of what was possible; anyone who had relatives in Westchester County or ventured to the mansions along Park Avenue, places of real money and real class, understood that.

"On a résumé, 1695 Grand Concourse looked good in the Bronx because the Grand Concourse loomed large on the internal cultural map of the borough," Corman said. "But it was not the leg up you'd imagine, especially during those years, with the anti-Semitism in the advertising business. In the eyes of the personnel guy, none of it was better. In the larger universe, they weren't hiring boys from the Bronx."

⇒ • ⇐

Like so many sons and daughters of the East Bronx, the novelist Jerome Charyn, born in 1937, has mostly miserable memories of his childhood. The son of barely literate immigrants, Charyn was raised on wild and unruly streets, "a garden of doom where nothing would grow except bitterness and regret," as he once described the place of his birth. Only once did he escape, and then only briefly. For a short time during the war, his family lived on Sheridan Avenue near 169th Street, in an apartment just down the hill from the Grand Concourse. Charyn arrived when he was five years old and left when he was seven. Yet this interlude might have been a lifetime, so deeply did it imprint itself on his consciousness, and so vividly does he coax this world to life in his three exquisite memoirs, most unforgettably in his 1997 work *The Dark Lady from Belorusse*.

The Dark Lady is largely a portrait of Charyn's mother, a Russian Jew called Faigele. As the dealer in a weekly poker game at the Concourse Plaza Hotel that is run by the Bronx's reigning power brokers, she comes to play a key role in the borough's black-market economy, caught up in a series of increasingly dramatic events.

Or something like that. Charyn's West Bronx, like the Odessa of Isaac Babel, the Russian storyteller to whose tales his work has been compared, is an otherworldly place, rich with imaginative flourishes. This world comes so vibrantly alive, both in these memoirs and in Charyn's essays and conversations about his childhood, that his West Bronx has the texture of a fever dream. The girls from the Theodore Roosevelt apartment complex are "so carefree, so unencumbered," the streets preternaturally safe day and night. Visual details catch the eye at every turning, notably the fanciful apartment houses named after the daughters, wives, and nieces of their builders—"the Beverly, the Diana, the Rosalind, the Sylvia, the Sandra, the Suzanne—that were like the sounds of sexy maidens who might leap from their own lettering on a stone façade and frolic with a boy of five." Most of all, his West Bronx is a world obsessed with language, words as evocative as those that the child everyone called Baby Charyn sought to master each month in the *Reader's Digest* quiz designed to help the magazine's audience improve its vocabulary.

The Grand Concourse moves through Charyn's memories like a royal highway. The boulevard is the preserve of Darcy the dentist, "a protégé of Bronx boss Ed Flynn, an underling who ruled the Grand Concourse like his own kingdom." Darcy is "that prince of the Grand Concourse," sometimes even "the king of the Grand Concourse," and the night the German surrender brought an end to the war, "the buildings basked in a silver light that was peculiar to the Concourse, as if sun and moon had met somewhere in the sky and were shining down on the West Bronx."

Abruptly as it began, this enchanted existence comes to an end. Back in the East Bronx, a forlorn place shorn of language, imagery, history, and much else, Charyn is reminded time and again of the gap between the two worlds of his childhood: "The kids in my class were so backward, they'd never been to the Concourse, and they'd never had art lessons, either.

I was the class whiz, who could talk about Bronx politics like no other boy. 'The Bronx reelected Roosevelt,' I said. . . . 'Boss Flynn and Mr. Lions managed his campaign from the Concourse Plaza.'" Yet bit by bit, he is reclaimed by the streets of his birth. "I had my own bed, and a little radio, and I could listen to 'Lux Presents Hollywood,' a condensed version of current Hollywood hits with lesser stars in the main roles . . . Tom Neal in 'Casablanca' and Barbara Britton in 'Madame Curie.' And slowly, slowly, the Concourse became a forgotten landscape, a lost article on a growing boy's map."

$$\equiv \quad \bullet \quad \equiv$$

Like Eliot Wagner, the poet Milton Kessler, the author of two incandescent works about the Grand Concourse, is barely remembered today; his admirers use the tactful word "underrated." Yet Kessler once loomed large on the landscape of American letters. For his ability to capture the lulling cadences of Bronx Jewish speech and give shape to the complicated circumstances of the urban Jew, he was credited for achieving in poetry what Bernard Malamud accomplished in the novel. The artist Ben Shahn created the frontispiece for one of Kessler's books, the poet Elizabeth Bishop wrote the preface for another, and the playwright Arthur Miller said of *Grand Concourse*, Kessler's fifth and final collection, published in 1990, "A good book like this is a gift to the world, a tourniquet to staunch the bleeding."

To Kessler, the Bronx boulevard was so beloved a muse that students of his work might suppose he entered the world on the street itself, perhaps at Dr. Leff's Maternity Hospital, a well-known medical facility located in a single-family house on the Grand Concourse near the Lewis Morris Apartments. In fact, that was the birthplace of his wife, Sonia, who lived on the Grand Concourse at 161st Street, and largely through her and her family did Kessler, who was born in 1930 in Brooklyn and grew up on Mosholu Parkway, become familiar with the boulevard and its varied moods.

As a teenager, Kessler played hooky a great deal, spending so many hours writing poetry in the wilds of Van Cortlandt Park when he should have been attending classes at DeWitt Clinton High School that he was eventually expelled for truancy. Two decades later, after a brief detour as a buyer of ladies' suits and coats, he alighted at the State University of New York at Binghamton, because, as his widow explained, "they needed a poet." He taught at the university for thirty-five years, until his death in 2000.

Kessler's most famous student, the feminist scholar Camille Paglia, once described her favorite professor as a "visionary rabbi . . . a burly, robust man with the inner emotionality of the melancholy Jewish tradition," and something of those qualities can be seen in photographs taken during Kessler's later years. He gazes benignly at the world through eyes magnified by huge glasses, his balding head framed by a wispy white corona. It is easy to picture this man wandering the neighborhood he eulogized so feelingly in his work, perhaps lounging on a bench in Joyce Kilmer Park, the Sunday papers in a pile by his side, or lingering outside a West Bronx synagogue after High Holy Day services, utterly at home.

Kessler's affection for the borough permeates those of his works set in the Bronx; holidays are celebrated, desultory conversations are overheard, and children, the poet's beloved daughter among them, mature and flourish. Yet Kessler's Bronx is not an entirely idyllic place. Many of the characters who move through his poems suffer from physical and emotional maladies; some are wounded soldiers and concentration camp survivors; others are victims of the Holocaust, of blindness and brain tumors, of loneliness and madness and old age. They sicken and die, frequently alone.

This amalgam of warmth and misery permeates the title poem of *Grand Concourse*, a series of twenty-two brief vignettes about family life, friends and enemies, the rhythms of the day, the passing of the years. Although the boulevard is never mentioned by name, a reader can imagine its small dramas unfolding on the streets of the West Bronx, in the bedrooms of its Art Deco apartment houses:

> The best times are at night when we sit
> in our bathrobes, read and talk.

in their sunlit living rooms:

> I see you're dressed up today, she said,
> No, these are my usual clothes, he said.
> I tucked my shirt in, that's all.
> It's the way I am. I just have
> a dressed-up look.
>
> A dressed-up look, she said, leaving.
> See you later, he said, smiling.

and on the boulevard itself:

> Holidays coming on. Her 87th birthday passes. No presents or visits,
> except for Doris, on the 3rd floor, who takes her shopping,
> sits with her in the park and has a bitter, miserly husband.

In "Mover," the one work in the collection in which the boulevard is mentioned by name, the street offers its own form of redemption:

> During the night a mover came
> And lifted
>   lightly lightly
> our whole apartment building
> and all our broken glass-wings
> and carried us
>   lightly lightly
> up to the Grand Concourse.

Like Chagall's dreamy cows floating above the rooftops of Vitebsk, Kessler's nondescript apartment house and its troubled souls are mysteriously transported through the air, then set down gently, to rest for eternity on the Bronx's street of dreams.

≡ • ≡

Even into the twenty-first century, the Grand Concourse continued to haunt the imagination of writers, among them Jacob M. Appel, a Bronx-born author of short fiction who revisited the boulevard's ghosts in a story entitled, simply, "The Grand Concourse," a haunting work that evokes the sense of desolation that by then clung so tenaciously to the street. The voice is that of a daughter who is escorting her troubled mother down "a lane of bad memories" that will lead to a devastating discovery. The boulevard's disrepair is a harbinger of the horrific discovery that awaits them:

> The Grand Concourse, once the Champs-Elysees of the Bronx, is now
> lined with Spanish-speaking travel agencies and discount tax preparers
> and storefront dentists advertising *no cash down*. . . . Placards announcing 24-hr BBQ or promoting "ultra-sizzlin" R&B albums seem out of
> place beside the Art Deco façades and creneled roofs unchanged

since the days of Joe DiMaggio. The neighborhood is thriving, yet as much a ruin as Petra or Pompeii.

A garrulous old drunk, mixing true local celebrities with those more tenuously linked to the neighborhood, offers to show them the sights: "You ladies wanna see where Uncle Miltie grew up? Where Steve Lawrence met Eydie Gorme?" The mother's old building, the Lewis Morris, has long since been stripped of its satin-gloved doormen and the crimson carpet on the sidewalk, yet intact is the closet that may or may not have been where the grandmother of the family wrapped a belt around her neck and hanged herself from a beam. Afterward, mother and daughter pass the old quarters of the Dollar Savings Bank, its admonitions to thrift overshadowed by come-ons for the smut shop next door ("Ask about our chocolate body topping"). The afternoon ends in violence, bloodshed, and profound sadness.

> We are the lead story on the local radio news: "In a bizarre incident this afternoon . . ."
> I drive aimlessly back toward the Concourse.
> The Bronx seems more itself at night. A light rain is falling and the streets are nearly deserted. It is possible to look down the boulevard and to imagine the Concourse in all of its lost glory. I can imagine my grandparents, arm in arm, wishing good evening to a white-gloved doorman and strolling up the street for a cup of coffee. But it is impossible to imagine what they are talking about, how they say it, who they were . . .

Appel's family moved to the affluent Westchester County suburb of Scarsdale when he was just a baby. Yet he grew up hearing endless stories about the Bronx and particularly about the Lewis Morris—recalled with such affection that, he said, "it was as if the Lewis Morris were a member of the family, perhaps a distant uncle." Around the time of his thirtieth birthday, Appel found himself drawn back to the Bronx, anxious to explore the terrain that had so profoundly shaped his family's history. Nevertheless, though he was fascinated by the place, and returned a dozen times, his characters find no solace in their past. In shocking fashion, they remind us that the worst of tragedies can occur even in what Appel's father used to call "the Jewish Buckingham Palace," that even crimson carpets and satin-gloved doormen offer no protection against heartbreak.

PART III

To Hell and Back

8

"The Borough of Abandonment"

EXECUTIVE TOWERS, the last luxury apartment house on the boulevard to be built with private money, opened at 165th Street in June 1963. With its wave-shaped balconies punctuating a curvy façade of glazed white brick, the twenty-three-story building looked like a vagabond from the Upper East Side come to settle in an unlikely yet still respectable part of town. Within weeks of the opening, many of the nearly 450 apartments were being dressed in flocked wallpaper, and deliverymen were carting mohair sofas edged with fringe into the spacious living rooms.

"It definitely represented a step up in the world," recalled a journalist named Barbara Graustark, who as a young girl lived nearby and used to visit friends in the building. "There was a sense of Executive Towers as the place where you had your first decorator. But you never had the feeling that this would be the forever place." Already, the Grand Concourse was a community in flux, and people were on the move, so much so that Executive Towers quickly became a mute symbol of the very decline it was intended to stave off.

At first the changes on and near the boulevard were small, sometimes barely noticed. The aroma of hard-to-identify foods drifted through the hallways of apartment houses, not the boiled chicken of countless Friday-night dinners in Jewish homes but more pungent smells that were new to these streets: rice and beans, or plantains. Graffiti appeared on the Lorelei fountain, then on the mammoth stone sculptures in front of the courthouse. Quiet nights were interrupted by the wail of police sirens or the rumble of fire trucks. Household garbage overflowed the trash baskets along the boulevard and spilled onto the sidewalk. In local newspapers, a growing number of articles were devoted to burglaries and robberies. Unfamiliar faces began appearing in elevators along the lower Grand Concourse and especially in buildings on the side streets—and by "unfamiliar" longtime residents meant black or Puerto Rican. Ever more often West Bronx residents were looking uneasily over their shoulders.

Soon, signs of change began coming faster. The local deli was held up at knifepoint. The owner of the candy store was held up, this time at gunpoint. Heroin produced bands of glassy-eyed kids who roamed the landscape, willing to do anything or hurt anyone in exchange for a fix; a decade or so later, crack cocaine would breed a new crop of predators.

The transformation of the southern half of the Bronx that began in the 1960s and continued for two decades is one of the most familiar stories in the history of the American city. The tale has been told again and again, in fiction and nonfiction, in film, in the art and music of the streets. Initially the causes seemed bewildering, even mysterious. In retrospect they were tragically obvious, born of the intersection of powerful social, economic, and political trends, along with profound generational shifts, and fueled by what proved to be disastrous policies on every level of government and in the private sector.

In recent years, an understanding of the forces that led to the decline of U.S. inner cities has grown exponentially and become far more nuanced. Scholarship of the late twentieth and early twenty-first centuries is demonstrating persuasively that the nation's urban crisis sprung from practices and attitudes with deep and twisted roots. To understand what happened to the Grand Concourse, and to the Bronx, it is necessary to look far into the past.

One historian who has explored this territory in detail is Ira Katznelson, a professor of history and political science at Columbia. In his book *When Affirmative Action Was White: An Untold History of Racial Inequality in Twentieth-Century America*, Katznelson examines how the social programs of Franklin Roosevelt's New Deal and Harry Truman's Fair Deal, ambitious though they were, helped widen the already considerable gap between whites and blacks. Katznelson describes how political deal-making ensured that domestics and agricultural workers, the job categories occupied by the great majority of southern blacks, were excluded from the protections of minimum-wage legislation, Social Security, unemployment insurance, and workmen's compensation. He also argues that the vaunted GI Bill of Rights, by being administered locally, denied black veterans the housing and business loans and especially the admission to white-only colleges that so boosted the fortunes of their white counterparts.

Thomas J. Sugrue, a professor of history and sociology at the University of Pennsylvania, has also traced the decline of U.S. cities to entrenched policies and practices, particularly those of the postwar period. In his book *The Origins of the Urban Crisis: Race and Inequality in Postwar Detroit*,

Sugrue scrutinizes patterns of public and private action and deep-seated assumptions that in many critical respects apply not just to Detroit but also to places such as the South Bronx. He points to endemic discrimination and exclusion that consigned blacks to the worst jobs, the worst housing, and the worst education. And he shows how this exclusion resulted from systematic actions in both the public and private sectors.

During these years, deindustrialization was remaking the fabric of American life and transforming American livelihoods. Hundreds of thousands of entry-level positions, especially the stable, well-paying, and mostly unionized factory jobs that had supported members of previous generations and allowed them to move up the economic ladder were fleeing the cities of the Northeast and the Midwest—to the suburbs, to the South, to the other side of the world—and in some cases vanishing entirely thanks to automation. In those years, money, population, and political power were increasingly concentrated in the nation's largely white suburbs, demonstrating that, as Sugrue sums it up, "to a great extent in postwar America, geography is destiny." Other students of the urban crisis have cited such factors as racial violence, white flight, and cultural forces that weakened already precarious family structures. But Sugrue argues that the forces he enumerates led to a systematic and entrenched racial and economic inequality, one that so limited the options of African Americans that overcoming barriers generally proved all but impossible, even for those who tried the hardest.

In the Bronx, the impact of these forces started making themselves felt after the Second World War, when waves of blacks from the Deep South, fleeing the loss of agricultural jobs and the residual constraints of a Jim Crow society, began descending on New York City, America's historical place of refuge. There they were joined by large numbers of Puerto Ricans seeking escape from the poverty of their island, a journey made easier by quick and cheap airplane flights.

Yet at the very moment these newcomers were arriving, the decline in suitable and available jobs was poised to interrupt the orderly flow of migration, a problem exacerbated by decisions being made two hundred miles away in the nation's capital. The vast latticework of federally financed highways that hastened the departure of these jobs was also transporting entire urban communities to the suburbs, where federally insured mortgages were making it ever easier to acquire a split-level dream house atop a pocket-size patch of green. Back in the cities, in the name of urban renewal, the federal government was leveling tenements in one poor neighborhood after another, replacing them with public housing projects that in turn displaced countless struggling and largely minority families.

In the West Bronx, the exodus to the suburbs was swelled by servicemen returning to the neighborhood and finding "no vacancy" signs on the rent-controlled apartments on and near the boulevard. Yet even had apartments been available, to the generation of West Bronx children who came of age after the war, nothing seemed attractive about the bourgeois streets of their youth. All those pricy oil paintings and overstuffed sofas that *Commentary* essayists Ruth Glazer and Isa Kapp had described so vividly had taken their toll, as had the sometimes tasteless materialism that they represented. Just as the parents of the Grand Concourse had fled the tenement neighborhoods of their own mothers and fathers, so their more sophisticated and affluent offspring couldn't escape fast enough from the fussy, overcrowded apartments in which they had been raised.

Even to their parents, the apartments seemed lacking. As the buildings aged, especially those along the side streets, the spaces within felt ever smaller and old-fashioned. Appliances and amenities that were new and exciting in their day had grown shabby with decades of use. Once-nurturing streets felt confining. And so it was not just grown sons and daughters who left but also families with young children. Year by year, then month by month, the West Bronx became increasingly the preserve of aging men and women who were reluctant and often unable to leave familiar and affordable quarters. The stagnation should have been a warning sign.

As one by one these elderly residents died or moved away, their apartments were snapped up by working-class blacks and Puerto Ricans from a South Bronx already bulging at the seams. The first families to arrive were almost as upwardly mobile as the Jews who had preceded them half a century earlier, and the initial reaction to the newcomers was, if not exactly welcoming, at least tolerant. "They're black but very nice," was the phrase commonly heard. Nevertheless, in an era in which black was coming to mean black power, clenched fists, and riots in the streets, when even the dark faces of Puerto Ricans provoked fear and antipathy among many whites, longtime residents grew ever warier. As they struggled to hold on to turf they regarded as rightfully theirs, their reaction to the newcomers grew noticeably less charitable.

Soon longtime residents were feeling uncharitable for other reasons. The first wave of new families set the scene for the arrival of others, many of them beset by a daunting array of problems—poverty, but also poverty's attendant woes, among them alcoholism, mental illness, family dysfunction, and especially drug addiction, fueled by the heroin that was making powerful inroads in New York City during the 1950s. Large numbers of these families were on welfare or close to it. By the mid-1960s, as the pace of change quickened, mobility among the indigenous population became flight, and by the end of the decade, the West Bronx had been transformed. Between 1940 and 1970, the percentage of whites dropped from 90 to 47, the percentage of blacks rose from 6 to 28, and the percentage of Puerto Ricans from 3 to 25. Under the weight of so many newcomers, arriving so quickly and in such distress, a social fabric was ripped, a social cohesiveness crushed. Among many of the whites who remained, the vestiges of once-entrenched liberalism were fast disappearing. Fear bled into a barely concealed racism that in turn morphed into panic.

The logarithm of neighborhood decay operated with a sickening predictability, underpinned by financial and governmental practices that, in retrospect, could only doom the community. Although the rent-control laws imposed during the Second World War remained in effect, the so-called vacancy decontrol law of 1971 allowed building owners to raise rents whenever an apartment changed hands, and landlords fast discovered the money to be made by rapid and frequent turnover of units. Welfare families, who were sometimes charged twice as much as their rent-controlled predecessors had been, represented an especially attractive proposition. The city offered finders' fees and other financial sweeteners to encourage landlords to accept these tenants and winkingly hinted that they would look the other way when it came to housing code violations.

With turnover quickening and families coming and going with increasing speed, landlords had ever less incentive to maintain their buildings and often let them decline with the complicit assistance of banks and insurance companies, neither of which had any great desire to support buildings in troubled neighborhoods and both of which had much to gain financially by running them into the ground. Though landlords complained, sometimes legitimately, of the difficulty of maintaining aging apartment houses with modest rent rolls, many landlords were at best indifferent and at worst corrupt. They abandoned buildings rather than pay taxes on them. They hired torches to set fire to their properties so they could collect the federally provided insurance they were required to have. They sold their buildings to speculators, who picked them so clean of wiring and appliances that the apartments were uninhabitable. The time it took for a once healthy building to sicken and die grew shorter and shorter. Although the sturdily built apartment houses directly on the Grand Concourse were largely immune from these forces, those on the side streets and the streets that ran parallel to the boulevard were not.

Perhaps the most draconian approach to the troubles engulfing neighborhoods such as the South Bronx was a strategy with the deceptively bland name of "planned shrinkage." The approach was articulated by Roger Starr, the city's housing commissioner, in a speech delivered in January 1976—"accelerate the draining" was the memorable phrase he used. Planned shrinkage was based on the argument that, given the city's current fiscal crisis, the most efficient way to deal with communities populated by the troubled and the troublesome was to starve them of essential services—police officers, firefighters, street cleaning, snow plowing—and by doing so force out the population that seemed to be causing all the difficulties.

Mayor Abraham Beame rejected Starr's recommendations, but the thinking behind them seeped into the national dialogue over how to respond to the problems of inner cities and from there into the municipal tool kit of approaches to urban disarray. In the southern half of the Bronx, services across the board declined, and during the 1970s, the borough lost 303,000 residents, all from the six community planning districts south of Fordham Road.

As the city abandoned these neighborhoods, they sank more deeply into chaos and disrepair. Garbage went uncollected, streets unswept. Like tumbling dominos, buildings and entire communities were turned into cauldrons of drugs, crime, arson, vandalism—every urban scourge imaginable. Block after block filled with abandoned husks of brick and concrete, soon to be replaced by empty lots in which waist-high weeds often represented the only living thing. The cycle was relentless and unstoppable. "Once abandonment begins in a neighborhood," concluded a 1977 report by the Women's City Club of New York on the changes buffeting the southern portion of the borough, "it spreads like a contagion from building to building, leaving wrecked, vandalized buildings and rubble-strewn vacant lots in its wake."

The blackout of July 1977 pounded the West Bronx, producing looting crowds that swarmed from one side of the Grand Concourse to the other, emptying and then trashing stores all along the side streets. But it was two events that autumn that marked the boulevard's low point, at least in the public eye.

On October 5, President Jimmy Carter's motorcade traveled up the Grand Concourse en route to the ruins of the South Bronx, a stop that made Charlotte Street, just a mile and a half east of the boulevard, the most famous slum in the country. And on a windy night exactly a week later, during a World Series game at Yankee Stadium, sportscaster Howard Cosell uttered to a national audience the phrase that came to stand as an epitaph for an era: "There it is, ladies and gentlemen, the Bronx is burning!" Cosell announced to sixty million people as an ABC cameraman in a helicopter panned to an abandoned elementary school in flames just a few blocks from the ballpark. By that point in the borough's skid downhill, flames exploding into a dark sky had become the single and most terrifying image of devastation. On the side streets branching off from the Grand Concourse and on the streets that ran parallel to the boulevard, fire decimated entire blocks. Lawyers who worked at the courthouse smelled the smoke from buildings that had burned the previous night when they got out of their cars the next morning.

An eloquent indication of the changing face of the Grand Concourse and its environs involved language. As recently as the mid-1960s, the term *South Bronx* referred simply to the cluster of communities at the borough's southern tip: Melrose, Mott Haven, Morrisania, the Hub. By the late 1960s, the South Bronx Model Cities district embraced a fifteen-thousand-acre area south of Crotona Park and east of Third Avenue. The northern boundary of what was considered the South Bronx edged up to the Cross Bronx Expressway, and by the end of the decade, everything below Fordham Road fell under that rubric. Nor did the watermark stop there. By 1985, an insurance agent was reminding a client who lived a few blocks north of Fordham Road that he too was officially a resident of the South Bronx.

The West Bronx was hardly alone in its travails. Other New York City neighborhoods suffered similar ravages during those years; central Brooklyn communities such as Bedford-Stuyvesant, Bushwick, and East New York were especially hard hit. Still, it is difficult to think of another part of the city, or even of the nation, that seemed so stable and then fell apart in such sudden and devastating fashion. By 1970, *New York* magazine had anointed the South Bronx "New York City's Number One disaster area," and the Bronx Realty Advisory Board christened the area "the borough of abandonment." A few years later, *Fortune* magazine described the South Bronx as "a social sinkhole in which civilization itself has all but vanished, . . . the closest men have yet come to creating hell on earth."

⇒ • ⇐

The elderly Jews who lived on and near the Grand Concourse were not the only victims of the crime engulfing the West Bronx. They were, however, among the most vulnerable, and given the racial sensibilities of the day and the tendency of local newspapers in those years to pay more attention to violence against whites than that against African Americans and Puerto Ricans, they were without question the most visible.

Newspaper articles gave these victims names and faces: Herman Nightingale, the sixty-eight-year-old retiree clubbed to death in his lobby at 1055 Grand Concourse, near 166th Street, at eleven on a Monday night as he was returning home from his regular late-evening stroll; Peter Friedland, the sixty-five-year-old pharmacist fatally shot after a struggle with two men who tried to hold up his drugstore on the boulevard near Fordham Road; the eighty-year-old woman raped and robbed during a thirty-minute rampage in her building at 1166 Grand Concourse during which another tenant, Esther Murciano, was robbed at knifepoint in the elevator. In 1969, when *Wall Street Journal* reporter Alan Adelson took the pulse of a single block along the Grand Concourse, just south of Mount Eden Avenue, he found a world transformed.

To describe this world, Adelson focused on an elderly resident named Tillie Halprin. "For Tillie Halprin," he wrote,

> life has become almost unbearable. Three times in the past four years, the frail, eighty-year-old widow has wakened to the horror of a burglar moving through the darkness of her bleak ground-floor apartment. Twice she was able to chase the intruder away by flailing at him with a cane and screaming. But not long ago, a burglar was too quick and too quiet for Tillie Halprin. She saw the dark silhouette for only an instant before a drugged cloth was pulled over her face. When she came to, the thief was gone—along with the $80 her son had sent her to help her get through another few weeks.

In almost all the families Adelson talked with, someone had been robbed, mugged, or assaulted within the past two years. The previous year, the number of felonies in the local precinct, the Forty-fourth, had risen 69 percent, the highest rate in the city and triple the citywide average.

Long before dark, residents barricaded themselves behind an arsenal of locks and bolts and chains inside apartments that had been transformed into fortresses. A school maintenance worker who lived around the corner from Tillie Halprin kept a collection of rifles, pistols, and steel pipes, along with cans of Mace, a hunter's bow and arrow, and a skin diver's speargun. His downstairs neighbor had a 165-pound Great Dane.

Routine outings became exercises in survival. A ride on the Grand Concourse bus, the journey that had so enchanted Bella Gross in Arthur Kober's innocent stories, was transformed into an obstacle course. Elderly women making their way up the steps confronted strapping youths sprawled on every other seat, their legs splayed, their expressions stony. Frail, fragile, and utterly intimidated, these women grabbed the railing near the driver, struggling to keep their balance as the bus lurched along the boulevard.

Within the space of five years, what the elderly men and women of the West Bronx once regarded as the Golden Ghetto had become a world they considered a jungle. Of the businesses on the block analyzed by the *Wall Street Journal*, only the local sales office for a burglar-alarm company was thriving. Fear was so rampant that when a woman suffered a heart attack while crossing the Grand Concourse, the local precinct was flooded with calls from neighbors saying they had heard she had been mugged. A few years later, in a landmark study of the Art Deco buildings of the West Bronx, Hunter College urban planners Donald Sullivan and Brian Danforth described the terrorization of the neighborhood's elderly by roaming teenage gangs and concluded, "At no time in the history of human civilization have we known such abandonment of the aged to violence." Earlier on, perhaps, these elderly men and women might have left. But by this point they were too poor, too fearful of the unknown, and convinced that life was no better anywhere else.

$$\Rrightarrow \bullet \Lleftarrow$$

To longtime elderly residents like Tillie Halprin, the newcomers to the West Bronx were simply anonymous perpetrators of crime and social breakdown—the torches, the heroin addicts and later the crack-cocaine addicts, the teenage welfare mothers giving birth to baby after baby, each with a different father.

Most of the newcomers did not fit the stereotypes. They struggled as much as anyone else—often much more. When they were sick, they waited hours in overcrowded clinics staffed with too few doctors. Their children went to school in broken-down buildings, presided over by harried and inexperienced teachers equipped with too few textbooks, and they played in parks filled with garbage and drug paraphernalia and used condoms. In the winter, their apartments had no heat and sometimes no water, and the broken locks on the front doors of the buildings made it easy for burglars to break in and for drug dealers to set up shop in the vestibule. When they worked late, they were mugged on their way home, and increasingly, they were mugged during the day. As ever more potent drugs tightened their grip on the streets, these new residents were often victimized by those closest to them. Most devastatingly, given their poverty, they had little if any means of escape, except to an even worse neighborhood.

Equally as much as longtime residents, they were confused and frightened by the changes around them. Yet as individuals, they were largely invisible. Not until years later did newspaper articles begin appearing about victims such as Maria Vidal Sanchez, a forty-nine-year-old Latino woman found slain in March 1978 in her ransacked apartment at 2922 Grand Concourse, near Bedford Park Boulevard.

Christopher Rhoades Dykema, a young social worker from the Midwest who worked in the West Bronx during those years, tended to many of these families. The physical disarray in which they lived was such that Dykema sometimes made his rounds with flashlight in hand because buildings had no lights in the hallways and trudged down blocks along which every structure had been abandoned and the city had stopped plowing the snow. The human toll was infinitely greater. "I used to see working-class parents ripped off by their own children," Dykema recalled. "I remember one particular mother on crack who couldn't remember where she had left her infant." Most vivid of all were the memories of the withdrawal babies, born to mothers strung out on narcotics. "You'd see them twitching," Dykema said, "and you'd know that their mothers were heroin addicts." The grandparents were typically hard-working African American or Puerto Rican couples, and while heartsick at the turn of events, they were the ones who ended up caring for the baby.

Some of the most tragic stories involved those who labored mightily to escape the troubled streets but with little success, among them some of the characters in *Random Family: Love, Drugs, Trouble, and Coming of Age in the Bronx*, Adrian Nicole LeBlanc's portrait of a collection of young people living in the 1980s on and near East Tremont Avenue, at the time one of the city's most drug-infested strips.

Their lives unfold on and off the Grand Concourse. They listen to the Rock Steady Crew at Poe Park; they sell pasteles at a bodega on the boulevard, and they punch numbers into their boyfriends' beepers from the boulevard's pay phones. Some of the players are drug kingpins, ruthless to family and strangers alike, but others yearn to wrest themselves from the violence and disarray that surrounds them. Particularly compelling is the character LeBlanc calls Coco, a streetwise but immensely appealing Puerto Rican girl who by her twenties is the mother of five children, one born with severe medical problems, by four different fathers. Coco tries heroically to raise her brood well, but every time she takes a step forward—by landing a job or going back to school—she is tugged back by the streets. Whether the obstacles she faces are personal or systemic seems less important than the fact that they are so overwhelming. It is easy to conclude that a young woman like Coco would have to be superhuman to break free and make a better life for herself.

One of the most detailed accounts of life in this environment is told by a onetime drug dealer named Allen Christopher Jones Jr.—the Rat, as he called himself in a memoir about his stormy coming of age near the Grand Concourse. The son of a West Indian mother and a father from Texas, Jones was born in 1950 in Lester Patterson Houses, a new public housing project a few blocks east of the boulevard. In its early years, Patterson Houses was a place of optimism for families who had moved there from tenement neighborhoods elsewhere in the city, a clean, safe, and attractive setting in which to grow up. But although Jones's childhood was happily uneventful—the city's projects had not yet begun their long downward slide—the rough streets beyond began beckoning.

Jones's life began skidding out of control in his early teens when he failed to make the basketball team at Taft High School, a devastating blow for an athletic, streetwise young black man. As Jones recalled,

> Shortly after I was cut from the Taft team, I began to hook up with a kid named John from 145th Street and Lenox Avenue in Harlem who also went to Taft. His cousin was a dealer and he was selling baby pounds of heroin, which usually sold for five dollars a bag. One day, when John saw me walking around the school looking down, which I definitely was after I knew I didn't make the team, he said to me, "Do you sniff?" When I said, "Yeah," he said, "Why don't you come to my crib," and I said "Bet."
>
> When we got to John's crib, he broke out a baby pound of heroin. We sniffed it slowly, and it seemed like all my troubles and insecurities were floating away on the high. Pretty soon, I was going to his crib every day after classes, until one day he said, "Do you want to try and sell some?" At this point, I had been sniffing for more than a year, so I said, "Bet, why not, I'll see what I can do."

Jones's first time out, he sold fifteen bags and collected fifty dollars. Soon he was shooting up in alleys on the Grand Concourse, so powerfully in the grip of the consort he calls "Bitch Queen Heroin" that "you would kill your own mother if she stood between you and your drugs." He strutted around the neighborhood wearing silk pants and a leather coat, a briefcase stuffed with cash under his arm, and his days and nights became a blur of selling, shooting up, and lurking outside banks and check-cashing places, knife in hand. Before Jones was out of his teens, he had been arrested and was sitting in a jail cell in Rikers, charged with five counts of armed robbery and possession of a deadly weapon, charges that could have sent him to prison for up to twenty-five years. After four months in Rikers, thanks largely to a sympathetic judge, a caring family, and an element of luck, he dodged hard time, an escape that allowed him to pursue a fifteen-year-long career in basketball, followed by a second career as a banker and radio commentator in Luxembourg.

⟹ • ⟸

Even during the worst of the violence, the Grand Concourse suffered less than some other parts of the Bronx did. The shrine that grew up on the site overlooking the boulevard where Joseph Vitolo claimed to have seen the Virgin Mary remained unharmed, as did the garden at Majestic Court, the building on the Grand Concourse at 196th Street where a longtime resident named Theresa Lato tended a little Eden planted with dogwood, mulberry, crabapple, and magnolia trees.

Starting in the 1980s, hip-hop culture draped the boulevard with a renegade glamour. Flavor Flav, clown prince of the sometimes incendiary rap group Public Enemy, lived for a time in Executive Towers and gained a measure of local celebrity when he was jailed for firing a shot at his next-door neighbor, who he claimed was having an affair with his wife. Beneath the boulevard, at the subway stop at 149th Street, graffiti virtuosos Fab Five Freddie, Daze, Crash, and their pals lounged on a worn wooden bench at the rear of the uptown platform, watching the gaudy Day-Glo rainbows they had created on the sides of the Number 2 and 5 trains race by in the darkness.

Because the solid Art Deco apartment houses along the boulevard were mostly spared the worst of the ravages, the Grand Concourse proper never resembled the burning Bronx that was so powerfully depicted by the news media. But the street hardly existed in isolation, especially the blocks south of Fordham Road. Charlotte Street, the official epicenter of disaster, lay just to the east of Crotona Park, and the Forty-fourth Precinct, which boasted the highest crime rate in the city, covered the neighborhood directly to the west. From many points along the boulevard, abandoned apartment houses and shells of buildings gutted by fire were visible as far as the eye could see.

Given the street's iconic role over the decades, the psychological impact of these changes were profound and profoundly disturbing. "The Grand Concourse had always been in part a street of imagination," Neil Sullivan points out in his history of Yankee Stadium. "It was as much a totem of aspiration as it was bricks and mortar." And so "when hard times hit the Bronx, the Grand Concourse was an easy, bitter symbol of dreams that had been shattered." By the mid-1970s, local residents were forced to acknowledge the fact that, as Sullivan put it, "the boulevard had become a major thoroughfare in a slum."

=== • ===

By 1975, Noonan Plaza in Highbridge was poised to become the first major Art Deco complex in the West Bronx to collapse. More and more welfare families were shuttled into the building where swans once swam amid water lilies in the courtyard pool, to be followed by squatters, vandals, and drug dealers. The landlord, who had owned the complex for less than five years, abandoned the place, leaving behind 327 housing code violations. "Noonan Plaza is being torn apart," wrote Donald Sullivan and Brian Danforth of Hunter College in their account of the decline. "Elevators, doors, windows, tiled ceilings, and garden furnishings have been brutally damaged. . . . Whole skylights have been taken off the roof and thrown to the ground below, sending broken glass in every direction, in order to sell the brass framing as scrap metal." Noonan Plaza, the authors concluded, "represents the worst that can happen to our fragile housing stock when it is left unprotected."

That year, an angel appeared in the form of the Settlement Housing Fund, an organization that had spearheaded the redevelopment of three thousand apartments for low-income and moderate-income families elsewhere in the city. When the fund committed itself to investing two million dollars to rehabilitate the building and convert its nearly three hundred apartments into co-operative units, salvation seemed at hand. "Noonan Plaza is an excellent example of what can be done to save our older buildings," Borough President Robert Abrams announced at a ceremony welcoming the arrival of a new boiler. A vice president of Manufacturers

Hanover, the bank involved in the reclamation effort, added optimistically, "In our experience, we have found that when people have a stake in their community, they work harder to maintain it."

By 1980, Horace Ginsbern's onetime showplace was in worse shape than ever. "Three years ago," a reporter named John Lewis wrote that year in the *Daily News,* "it appeared to be a dream come true for 210 families of Noonan Plaza in the Bronx when the Settlement Housing Fund took over and began rehabilitating the badly declining apartment complex. . . . Last week, all that was left was a nightmare for the remaining six families, who were told to pack up their belongings and leave."

Swamped by the challenge of rescuing such a large and troubled development, the fund had pulled out of the deal; the problems, admitted Clara Fox, the organization's director, were just "too overwhelming." Into the vacuum rushed a new crop of vandals who roamed the building freely and tore out so much of the plumbing that the water had to be shut off. Finally, citing a lack of basic services, the vacant apartments ransacked and filled with garbage, some four hundred violations, and "most parts of the building in total disrepair," the city ordered everybody out.

Lorenzo Dufau, a former tenant, watched the final chapter from the sidelines. "It is a disgusting feeling to know you have been used like that," Dufau told a reporter. "This place was beautiful, with parquet floors and big bathtubs that a man could get into. Now they are ripping out and destroying everything without any feeling. The good people who cared about this place were defeated in every way."

On the boulevard, the Lewis Morris was similarly on the ropes. The apartment house once so elegant that children were not allowed to roller-skate in front of the building lest the noise disturb residents had, like Noonan Plaza, became a dumping ground for welfare recipients. The entire staff was let go. In the lobby, leaks left holes in the ceiling, crumbling plaster carpeted the marble floor, and nests of shredded plaster remained where slabs of marble had been stripped from the walls. "There's no doubt that Lewis Morris will burn," a city building inspector told a graduate student doing research on the building. "How can we stop it? Impossible, you can't prevent a fire before one is intentionally set. You can't press violations against a landlord you can't find. You can just condemn the building and get the tenants out and hope the fire doesn't hit too soon." It was a shame, the inspector added. "When I was a kid," he said, "everyone I knew wanted to live there."

⇒ • ⇐

Of all the apartment houses along the Grand Concourse, no place suffered more than the palatial complex that half a century earlier had helped usher Logan Billingsley into the ranks of upper-crust Bronx society. In the 1930s, the Theodore Roosevelt was rechristened Roosevelt Gardens and underwent a facelift that removed all traces of the Spanish-accented décor. Yet not until the late 1950s, when the family of Wolf Weinreb bought the buildings, did Roosevelt Gardens begin its long sad march toward collapse.

In the mid-1970s, a young out-of-work economist named Gelvin Stevenson, who lived nearby, used to pass Roosevelt Gardens during his walks around the neighborhood and was especially struck by the sight of tenants filling buckets with water from the fire hydrant on the street and bringing them back to their apartments. Being familiar with the financial underpinnings of residential development, and baffled as to how a once-solid project could disintegrate so quickly, he began examining the situation more closely.

Stevenson's discoveries, described in an article entitled "The Abandonment of Roosevelt Gardens" that was published in 1979 in *Devastation/Resurrection: The South Bronx*, an exhibition catalogue of the Bronx Museum of the Arts, served as a parable for the downfall of the entire West Bronx. His account is critical to an understanding of the changes that engulfed the area because it provides so detailed a look at exactly how a building dies and, more important, what life was like for the largely African American and Latino families whose home literally fell apart before their eyes.

During the fourteen years that the Weinrebs owned Roosevelt Gardens, the buildings bounced back and forth in a series of increasingly shady transactions, presumably for tax benefits. The family sold and repurchased the complex so frequently and so rapidly that it was sometimes sold and bought back on a single day.

Deterioration started worsening in about 1973 as apartments began filling up with families from the South Bronx left homeless by fire and building abandonment. At the point that Stevenson obtained the rent rolls, the complex was occupied almost entirely by minority women with large families—often six, seven, or eight children—and by year's end, maintenance had virtually stopped. The basement flooded around Christmas, and "when the phone company came to repair the lines," one tenant told Stevenson, "they refused to enter the basement because there were so many rats and water bugs."

As badly as Roosevelt Gardens was beset by corruption and neglect, few tenants were prepared for the horrific event that took place on October 4 of the following year involving an eight-year-old black girl named Angelia Holden—her friends called her Missy—who lived with her mother and five siblings in a sixth-floor apartment. At three that afternoon, Missy got into the elevator in the lobby with a friend and pressed the button for her floor. A report subsequently filed with the city buildings department, based on an account from a police officer who had been called to the scene, described what happened next: "Detective George Pagano . . . told me Angelia Holden . . . attempted to leave elevator through vision glass on cab door which was missing. . . . Both girls had entered elevator and the shaftway door and cab door closed and elevator did not start. Angelia tried to get out of elevator when elevator started up and she was crushed, 'head' between car and elevator shaft." Apparently as Missy Holden started to crawl through the broken window in the cab, the elevator began moving up, crushing her skull. She was pronounced dead on arrival at Morrisania Hospital. Asked about the matter three years later, Weinreb told Stevenson, "The insurance took care of that."

In June 1975, Roosevelt Gardens was sold for the last time, to a man named David Teichner who paid eleven thousand dollars for a complex then assessed by the city for tax purposes at $1.3 million. Teichner was what was known as a finisher. He stripped the place of all its salvageable materials—refrigerators, stoves, toilets, brass pipes, some of which were yanked out even as water was running through them—then sold them for profit.

With the water turned off because of flooding, tenants brought empty buckets down to the fire hydrant in front of the building, filled them to the brim, and, because the elevators were by now broken, carried them upstairs to their apartments so they could use the water to cook and bathe and flush their toilets; this was the ritual that Stevenson had witnessed on his strolls around the neighborhood. Fires were breaking out at the rate of three or four a day; the fear of fire was so great that, one tenant told Stevenson, "we went to bed with our shoes on every night." The day that Seymour Posner, the local state assemblyman, came to pay a visit with his staff, he claimed to have found more than a thousand violations in the public areas alone.

Later that year, the stucco palace where dogwoods once bloomed and references were needed to rent an apartment became the first building on the Grand Concourse to be abandoned. By October 13, when the city ordered out the remaining twenty-eight families, nearly all the windows had been smashed, and rubbish blanketed the once exquisite garden. For the next three years, the complex stood empty, a looming, ghostly presence, its windows covered with blue sheet metal and the entrance and the courtyard sealed off with an eleven-foot-high fence of chicken wire. At one point, wild dogs roamed the premises.

Roosevelt Gardens became the ultimate cautionary tale. "Don't let Roosevelt Gardens happen to your building," Assemblyman Posner warned in a newsletter to his constituents. Two years later, when Stevenson pressed Weinreb about the financial transactions that had brought down the complex, the answers he received were evasive. "Don't have old files, don't have nothing," the former owner said. "Building's all gone now. I don't know anything about it now." Hadn't Weinreb kept records for tax purposes? "We only hold what is necessary," he replied.

≡ • ≡

Ellen Samuels, who was born near the Grand Concourse in 1926 and left the neighborhood when she was sixteen years old, tells a familiar story about how change arrived unawares. Like virtually every child of the West Bronx, she had vivid memories of the social events held at the Concourse Plaza Hotel, and in 1970, when she was planning a celebration for her parents' fiftieth wedding anniversary, the hotel seemed the ideal setting. "The Concourse Plaza had been so nice," Samuels recalled. "All those affairs." But she had not set foot in the Bronx for thirty years. When the time came to check out the place in preparation for the party, she discovered that the Concourse Plaza had become a welfare hotel.

Drugs and crime defined the hotel's rhythms, with the staff as much at risk as the tenants, as the hotel's manager discovered one August morning in 1971. According to the police, the manager had gone to a resident's room the previous evening after neighbors complained about noise and found a man beating a woman. The man left the next morning but returned a few minutes later to demand the two-dollar deposit for his room key. In the scuffle that followed, the man pulled out a 32-caliber pistol and fired three shots. One of them fatally punctured the manager's chest, and another entered the stomach of the eighty-year-old bellhop who was trying to stop the assailant with a chair.

In 1974, as a renovated hundred-million-dollar-plus Yankee Stadium was rising a few blocks to the west, the city took over the hotel, with plans to convert it into housing for the elderly and use the ballrooms for municipal offices that, tellingly, had been crowded out of the Bronx County Courthouse to make room for new narcotics courts. Although the hotel had been emptied of its residential tenants, the premises were a mess. The crystal chandeliers, miraculously still swinging from the ballroom ceilings, had turned silvery with dust. Civil servants worked amid drafts and flooding in spaces whose signs identified them as the Wedgwood Room and the Crystal Room. The crimson carpets were threadbare, and visitors with such names as Ace 150 and Rap 164 had left their calling cards on the gilt wallpaper. The grass-green carpet in the bedrooms crunched underfoot, so thick was the layer of pink plaster that had fallen in chunks from the ceiling. In an upstairs office, in a space stinking of rotten cat food, newborn kittens huddled near a cloakroom under a sign reading "Not responsible for fur coats or valuables."

Four years later, with public development around the city at a standstill due to the city's fiscal crisis, the plan to rescue the hotel was no closer to completion. Even the mobsters who once gathered in the lobby after Yankees games had abandoned the place. By the summer of 1979, director John Cassavetes was filming the opening scenes of his movie *Gloria*, starring his wife, Gena Rowlands, in the deserted lobby, and so transformed were the premises that no one who had known the hotel in its heyday would have recognized the images they saw on screen. The space appears as a Piranesi landscape of shadowy and forbidding stairways that end in pools of darkness. Only the cats continued to thrive. "There must have been six thousand cats in the lobby when we first got there," Rowlands told an interviewer.

≡ • ≡

Halfway up the boulevard, at Loew's Paradise, change began to arrive as early as the 1940s, when the orchestra pit was covered with concrete to create additional space for moviegoers. In subsequent years, the red velour seats were reupholstered again and again, each time with more durable fabric, and the lobby walls were stripped of their salmon-colored silk and painted a color called Loews Buff. Yards of climbing vines and countless artificial birds were removed from the walls, paintings and statues disappeared one by one, and the asbestos curtain adorned with a full-size image of a Venetian garden gave way to a more utilitarian number decorated with what one observer described as "a rather melancholy single tree."

On the façade, the Seth Thomas clock ran down, and around 1970, the statue of Saint George "mysteriously disappeared from astride his charger," according to theater historian Michael Miller, "apparently lowered five stories in full view of the busiest street in the Bronx." The dragon had also been stolen, and the Wonder Organ eventually ended up in New Jersey. Although high school graduations were held in the auditorium into the 1970s, the building's days as a playhouse were numbered. In 1973, the theater reopened as the Paradise Twins, and a decade later the interior was sheathed in wallboard to ease its transition into a fourplex. The stars had long since stopped twinkling, clouds no longer drifted in the sky, and the goldfish had passed on decades earlier, victims of too much Coke, too much popcorn, and slingshots fashioned from rubber bands and paper clips. By the time the Paradise was shuttered in 1994, virtually every trace of what Drew Eberson, the architect's son, described as his father's "wonderful bits of showmanship" had disappeared.

⇒ • ⇐

At the Andrew Freedman Home, paper plates had by the 1960s replaced bone china, and accommodations came with a price tag. When journalist Vivian Gornick paid a visit in 1980 on behalf of the *Village Voice*, barbed wire surrounded the building. Two years later, a community group called the Mid-Bronx Senior Citizens Council bought the home, with plans to convert "the ex-rich men's poorhouse" into subsidized housing for the elderly poor. As its fortunes declined, longtime handyman Pete Pitta kept the library locked so people wouldn't steal the books.

Synagogues up and down the boulevard, long unable to summon a minyan, were repurposed and renamed. The luckier ones loaded their Torahs and bronze plaques onto buses and carted them up the Grand Concourse to houses of worship in Riverdale or Westchester County, where these institutions lived on largely in memory; others simply disbanded. Vandals stole the statue of the raven at Poe Cottage and tried to set the place on fire; by the late 1980s, crack-cocaine dealers had taken over the onetime "little green jewel." Alexander's closed abruptly one May day in 1992, and the cafeteria where Joey Hacken had held court gave way to an offtrack betting parlor, though this transformation proved less traumatic than many others as the bookies who hung out at Bickford's simply set up shop in the new place.

⇒ • ⇐

The Lorelei fountain, a victim of vandalism in the 1970s, emerged as one of the most potent expressions of the neighborhood's troubles during those years. (*New York Times*)

Louis Risse's "avenue of pleasure" was almost unrecognizable. With the quickening exodus of the Jews of the West Bronx, High Holy Day promenades along the boulevard had gone the way of the well-brushed Homburg, and smiling political candidates atop open convertibles had become a faint memory. Many of the trees along the boulevard had long since been relocated to Pelham Bay Park or had died through neglect, and the little memorial plaques honoring the fallen soldiers of the First World War had been stolen by drug addicts. The boulevard of old enjoyed a brief star turn on October 28, 1974, when television's Rhoda Morgenstern, late to her own wedding, raced across all eleven lanes of the street as she tried not to trip over the train of her white gown en route to her parents' Grand Concourse apartment. But mostly it was business as usual, even as local officials sought to recapture elements of the street's former glory. According to the *New York Times*, in the predawn hours of the Sunday afternoon in June that people were celebrating what the city fathers were calling "Bronx Week 1988," the police raided a crack factory at 1814 Grand Concourse near 175th Street, where they seized eight handguns, three shotguns, two Uzi machine guns, and a homemade pipe bomb, along with a large cache of narcotics.

The marble mermaids of the Lorelei fountain in Joyce Kilmer Park, a particular target of the rampaging of the 1977 blackout, were coated with graffiti and stripped of their heads and arms and fins in vandalism so brutal that one city official recommended moving the sculpture to a neighborhood whose residents would, as he put it, "better appreciate" its charms.

Large green and white exit signs reminiscent of those found on the interstates sprouted at even minor intersections along the Grand Concourse, a tacit acknowledgment that the boulevard had devolved into a high-speed traffic artery. Its roadways were widened, and medians were narrowed or removed entirely. At one point, instead of planting grass in a median, the city filled it with green concrete.

With traffic came traffic-related fatalities. By 1989, an average of one pedestrian was being killed on the roadway every month, a pace that made the Grand Concourse second only to the notorious Queens Boulevard among what the tabloids called New York City's "boulevards of death." Three years later, even as pedestrian fatalities declined citywide, officials were describing the Grand Concourse as New York's most dangerous street. "The Grand Concourse has become what none of us want: a highway," Borough President Fernando Ferrer announced glumly after a five-year-old was killed and a two-year-old was hit by a truck.

By the 1980s, when Tom Wolfe was furiously churning out chapter after chapter for *Rolling Stone* magazine of what would become *The Bonfire of the Vanities*, his acid-drenched portrait of the late-twentieth-century city, the intersection of 161st Street and the Grand Concourse stood as the epicenter of a metropolis spun out of control. Assistant district attorney Lawrence Kramer, one of Wolfe's vividly drawn cast of characters, watched the downward spiral from his office in the great granite courthouse, reminded every day that this world had emerged from the ashes of his father's Bronx, "the old Bronx in all its glory."

> Up there at the top of the hill, 161st Street and the Grand Concourse had been the summit of the Jewish dream, of the new Canaan, the new Jewish borough of New York, the Bronx! Kramer's father had grown up seventeen blocks from here, on 178th Street—and he had dreamed of nothing in this world more glorious than having an apartment . . . someday . . . in one of these grand buildings on the summit, on the Grand Concourse. . . . Did you want an apartment on the Grand Concourse? Today you could have your pick. The Grand Hotel of the

Jewish dream was now a welfare hotel, and the Bronx, the Promised Land, was seventy percent black and Puerto Rican.

The courthouse still stirred Kramer's soul, but the building had come to represent as much a fortress as the apartments in which the neighborhood's elderly Jews barricaded themselves at day's end.

> You could ascend to the very top of the criminal justice system in the Bronx and eat deli sandwiches for lunch until the day you retired or died.
>
> And why? Because they, the Power, the Power that ran the Bronx, were terrified! They were terrified to go out into the heart of the Bronx at high noon and have lunch in a restaurant! Terrified! . . . The heart of the Bronx was now such a slum there was no longer anything even resembling a businessman's sit-down restaurant. But even if there were, what judge or D.A. or assistant D.A., what court officer, even packing a .38, would leave Gibraltar at lunchtime to get it?

⇒ • ⇐

At Philly's, the candy store just off the Grand Concourse near 165th Street, the Barishes watched as the tailor next door was held up, and then the grocery store was. Rita Barish started getting to the store at six thirty in the morning so as not to leave her husband there alone. In those years, Philly's went through three separate sets of iron gates. In the first break-in, intruders took only cigarettes and a little cash. The second and third times, Rita walked in to find only a shell of a store, thousands of dollars of merchandise gone. After that, when she saw a strange face, her heart pounded.

The situation was no better around the corner at 215 East 164th Street, the building that had been the couple's home for three decades. At six p.m. on one Memorial Day in the late 1960s, Jack Cohen, the awning-store owner, was stabbed to death in the lobby as he waited for the elevator. That weekend, the Barishes decided that the time had come for them to leave also.

On August 1, 1976, twenty-five years and two months after he had bought the store, Phil died of a heart attack while getting dressed for work. Two weeks later, Rita sold the business, and not until a winter day in the early 1980s, driven by a mixture of sentiment and curiosity, did she venture back to the old neighborhood. Stepping gingerly over the garbage that lay in her path, she braced herself for the worst. Even so, she was not prepared for what she found. The bright and bustling place she had presided over for a quarter of a century seemed dark and shabby and forlorn.

Hers was the only white face. To cover her confusion, she made a quick phone call, then fled.

Back in her high-rise apartment in New Jersey, on the other side of the Hudson River, someone mentioned egg creams. "I don't know if they know how to make a good one," she replied sadly. The truth was, the Puerto Rican woman who had taken her place behind the counter had not been asked to make an egg cream in years.

⇒ • ⇐

Starting in the mid-1960s, the growing woes of the Grand Concourse and its environs prompted a blizzard of reports by public and private agencies, fat documents brimming with dismal statistics and concluding with impassioned calls to action. The patient is gravely ill, the studies warned, and the wounds must be staunched before it is too late. Nevertheless, to read these documents, especially in retrospect and one after another, is an exercise in heartbreak. Though written with the best of intentions, often by individuals possessed of intelligence and idealism, these studies did not acknowledge, nor, perhaps, could their authors have realized, that the moment for rescue had long since passed.

Even the city's official master plan, an ambitious six-volume work published in 1969, arrived with a thud. "The Master Plan was out of date the moment it was printed," the *New York Times*'s longtime architecture critic Ada Louise Huxtable wrote nearly two decades later. "The city had turned out to be a fragile and interdependent, highly unpredictable, fractious, erratically functioning entity far beyond planners' conventional ideas or projections and often in conflict with them." Efforts to eradicate blight by making physical improvements, Huxtable continued, "had failed to alleviate the pressures of poverty and deprivation or the growing cataclysm of drugs and despair."

Many West Bronx residents sensed the irrelevance of large-scale physical solutions to entrenched social and economic problems even as the shiny, oversized volumes—bright purple for the Bronx—were being delivered to local libraries. As one resident told city planning commissioners who trooped up to Taft High School for a public hearing on recommendations for Community Planning District Four, which covered the Concourse-Highbridge neighborhood, "I want to make clear I don't have much faith in the whole planning process."

The flurry of government programs designed to reverse or at least slow the pace of decay seemed equally beside the point. By the late 1960s, small-bore solutions such as bookmobiles, wading pools, police scooters, and bright orange street lights struck local residents as almost quaint. Given the gravity of the situation, perhaps the best of efforts might have failed. "Change was so dramatic in the Bronx," said Gary Hermalyn, executive director of the Bronx County Historical Society, "one could easily argue that ten times the effort might not have changed the results one bit."

⇒ • ⇐

Ruth Puttermesser, the erudite heroine of Cynthia Ozick's 1997 novel *The Puttermesser Papers*, watched the decline of the Bronx from her childhood apartment on the Grand Concourse.

> Puttermesser was thirty-four, a lawyer. She was also something of a feminist, not crazy, but she resented having "Miss" put in front of her name. . . . Though she was no virgin, she lived alone, but idiosyncratically—in the Bronx, on the Grand Concourse, among other people's decaying old parents. . . . She roamed the same endlessly mazy apartment she had grown up in, her aging piano sheets still on top of the upright with the teacher's X marks on them showing where she should practice up to.

Puttermesser's life takes many strange turns. After leaving a Wall Street firm and going to work for an obscure city agency, she is visited by a golem, a mythical figure from Jewish folklore, with whose help she becomes the city's most benevolent mayor. Lovers come and go. At the age of ninety she dies, the victim of rape and murder, only to wind up in Paradise, albeit a heaven of dubious delights. The event that seems to set in motion all the drama of her life is the destruction by vandals and addicts of the world of her childhood, an act that comes to symbolize the growing chaos of contemporary urban existence:

> Her majestic apartment on the Grand Concourse in the Bronx, with its Alhambra spaciousness, had been ravaged by arsonists. Even before that, the old tenants had been dying off or moving away, one by one; junkies stole in, filling empty corridors with bloodstained newspapers, smashed bottles, dead matches in random rows like beetle tracks. On a summer evening Puttermesser arrived home from her office without possessions: her shoes were ash, her piano was ash, her piano teacher's penciled "Excellent," written in fine large letters at the top of "Humoresque" and right across the opening phrase of "Für Elise," had vanished among the cinders.

Who Killed the Concourse?

PEOPLE LOOKING FOR SOMEONE or something to blame for the tumultuous changes washing over the Grand Concourse didn't have to look far. Co-op City, the cluster of brick towers on the great sweep of marshland along the Hutchinson River, emerged as so potent a symbol of the dismal fortunes of Louis Risse's great boulevard that decades' worth of motorists crawling along the Hutchinson River Parkway found it impossible to pass the huge apartment complex off to the west without the thought crossing their minds, "There it is. That's what killed the Grand Concourse."

Although Co-op City has the immutable look of some urban Stonehenge, a fixture on the landscape seemingly forever, the complex was not the first structure to occupy the site. Before Co-op City there was Freedomland, a 205-acre amusement park that during its brief and unhappy life billed itself as the world's largest entertainment center.

Freedomland, which was shaped like a giant map of the United States, offered a number of beguiling attractions. Visitors could ride through authentic Civil War gunfire, wave at Elsie the Cow, admire gentle renditions of the nation's topography—Rocky Mountains that rose to a modest fifty feet, Great Lakes just seven feet deep—and, in an ironic tip of the hat to the city's low crime rate during those years, watch faux robbers hold up a bank in what was styled "Little Old New York." On opening day in June 1960, visitors were so eager to experience what the New York Times described as history with the "dull places" left out that traffic was bumper to bumper on the roads leading to the site, and families started arriving at seven thirty in the morning, two and a half hours before pop singer Pat Boone cut the ceremonial ribbon.

Yet despite the park's earnest aspirations, not to mention a catchy jingle that lingered in the minds of New Yorkers long after the park itself was just a memory, the expected crowds never materialized, and Freedomland proved a bust. The mosquitoes didn't help, nor did the 1964 world's fair, which arrived four years later on the other side of the East River in Flushing Meadows Park. Perhaps most fatally, Freedomland lacked the element

of fantasy that made Disneyland in California such an instant and enduring crowd-pleaser. William Zeckendorf, president of the company that
owned the land, acknowledged as much when he described the undertaking as "misconceived, grievously mislocated, and utterly mismanaged." In
September 1964, Freedomland filed for bankruptcy, an act that cleared the
way for a megaproject that many people came to regard as an even more
disastrous undertaking.

=== • ===

But not right away. Anticipation about Co-op City's opening ran so high
that the day after the project was announced in February 1965, 1,698 families showed up at the offices of the United Housing Foundation, the spon-
soring organization, to put their names on the list for apartments. The first
edition of the *Co-op City Times* appeared in October 1966, more than two
years before the first moving van arrived, and even at that early juncture,
the publication's editors sensed a moment touched by destiny. "For those
with an interest in history," one breathless article predicted, "we believe
the early editions of this newspaper will someday be of value and interest
in tracing the development of building a new city."

Co-op City sprung from the same strain of idealism that four decades
earlier had given birth to Andrew Thomas's namesake apartments on the
Grand Concourse. The project's spiritual father, a Ukrainian immigrant
named Abraham Kazan, had been a pioneer in the movement to provide
affordable housing for families of moderate means, and the United Housing Foundation, the union-supported organization he created in 1951, was
the country's leading developer of co-operative housing.

Like Thomas, Kazan grew up with firsthand knowledge of the miseries of
tenement life, and also like Thomas he believed passionately in the importance of attractive apartments for families with limited incomes. "We are not
only demolishing rat-traps and building decent housing," Kazan used to tell
his critics. "We are giving people the opportunity to let the sun into their
apartments and enjoy gardens and trees." His first co-operative venture,
Amalgamated Houses in the North Bronx, which he founded and ran for
forty years, represented a patch of heaven on earth to the left-leaning residents who moved there in 1927. And in a lovely touch of symmetry, the architect chosen to design Co-op City, the Russian-born Herman Jessor, had
been a junior member of the Amalgamated's design team.

On the marshy acres west of Pelham Bay Park, these high-minded
men seemed poised to create something of great scale and social

usefulness—housing for up to sixty thousand people earning between seven thousand and seventy-five hundred dollars a year. Purchase prices and monthly fees would be kept low thanks to the state's Mitchell-Lama program, which subsidized housing for middle-income families.

Unlike the Great Wall of China, Co-op City could not be seen from the moon. Its reputation on Earth, however, was considerable. Routinely described as the largest single apartment development in the country and the largest co-operative community in the world, the project cost $314 million to build and involved numbers so huge that they often fluctuated in the telling. Plans for the 320-acre site called for thirty-five towers and 236 townhouses providing a total of 15,372 apartments, along with three shopping centers, half a dozen public schools, and space to park 10,500 cars. Some forty-one city agencies were involved even before the first contract was let; the agreements covering the creation of the street system alone ran to a hundred pages.

Although the United Housing Foundation was run by savvy, hardheaded individuals who had built major developments throughout the city, including the high-profile Penn Station South, an enclave of garment workers in the heart of Chelsea, the group's project in the northeast Bronx was clothed in appealing rhetoric and imagery, even down to the emblem of twin pine trees that was its logo. One brochure written to introduce prospective co-operators to their new home described Co-op City as "friendly people living together," and something of the early euphoria shines out of a photograph taken on May 14, 1966, to commemorate the ground-breaking. Aging, snowy-haired lions such as Kazan and Robert Moses stand shoulder to shoulder with members of the next generation of political figures, among them Governor Nelson Rockefeller and Borough President Herman Badillo—significantly, the first Puerto Rican to occupy that position—wearing dazzling smiles and looking vigorous enough to erect one of those brick buildings with their own hands. A rainbow assortment of children holding small ceremonial shovels adds to the sense of occasion.

The warm feelings faded the moment the first tower rose from the swampland, washed away by a torrent of almost uniformly bad press. Although Ada Louise Huxtable of the *Times* praised the "good apartments at unbeatable prices"—two bedrooms and a terrace could be had for $2,250, with monthly payments of $129—most critics loathed the place. *New York* magazine described the complex as "fairly hideous," and *Time* magazine judged its towers "overbearing bullies of concrete and brick," its layout

The 1966 ground-breaking for Co-op City, the huge co-operative housing development in the northeast Bronx that attracted so many West Bronx families, was attended by master builder Robert Moses, Governor Nelson Rockefeller, co-operative pioneer Abraham Kazan, and Borough President Herman Badillo. (Sam Reiss)

"dreary and unimaginative," and the total effect "relentlessly ugly." Most devastating, however, was the charge that Co-op City, by its mere appearance on the scene, had single-handedly destroyed the Grand Concourse and by extension the entire West Bronx by providing an escape route for white residents desperate to flee a declining and racially changing part of the city.

Exact numbers proved unexpectedly elusive. According to some newspaper accounts, the United Housing Foundation acknowledged that up to four thousand of the first ten thousand applicants were white families from the heart of the Bronx, many of them from the boulevard proper, although the organization's definition of exactly what constituted the West Bronx was vague and by one reckoning extended all the way to Prospect Avenue, a street more than a mile east of the Grand Concourse. The Concourse Jewish Community Council estimated that between ten thousand

and fifteen thousand West Bronx residents moved to Co-op City as soon as apartments became available.

But even the most conservative estimates, such as one from the American Jewish Congress that only 15 percent of Co-op City applicants came from what the organization described as the Grand Concourse area, could not mask the fact that huge numbers of West Bronx residents were heading for the brick towers rising in the east. The drumbeat of newspaper articles forecasting further deterioration of the West Bronx due to the arrival of Co-op City only exacerbated local fears, despite the fact that the departure of large numbers of West Bronx families had begun a decade before the project arrived.

On March 9, 1967, under the headline "City Fights White Exodus from Grand Concourse," Abel Silver, a reporter for the *New York Post*, forecast the rapid demise of the once-flourishing boulevard. Claiming that 80 percent of the first applications for apartments in Co-op City came from residents of what was defined as the Grand Concourse area, Silver predicted that the opening of the complex "will undoubtedly hasten the Grand Concourse area's decline." And he added, "Worried city officials are desperately seeking ways of maintaining a stable, integrated community."

Coming from the liberal newspaper that was something of a bible in the heavily Jewish West Bronx, Silver's words had enormous impact, and his grim predictions were repeated over and over. "Some critics [assert that] Co-op City will help to create future ghettos by 'siphoning off' middle-class white families from Bronx areas that are in racial transition," *New York Times* housing reporter Joseph Fried wrote in March 1968. By 1975, *Fortune* magazine had concluded, "It is both ironic and symbolic that the most telling blow of all to the stability of the South Bronx was struck by do-gooders." Within a year after the first tenants arrived in December 1968, the magazine noted pointedly, the Bronx began to burn.

⇒ • ⇐

The conventional wisdom was that the West Bronx fled to Co-op City exclusively to escape escalating crime and the growing number of dark and unfamiliar faces. The reality was more complicated. For early residents such as Helen and Sol Schwartz from Morris Heights, a community just west of the Grand Concourse and north of the Cross Bronx Expressway, the spanking new apartments were as strong a magnet as the yearning to depart the old neighborhood.

In many respects, the couple lived an utterly typical West Bronx life. Helen Schwartz, the daughter of a house painter, grew up in the East Bronx—on Kelly Street, among other addresses. Her future husband, who sold fur coats, lived next door. In 1945, the year Helen turned twenty, they married and moved to a one-bedroom on the top floor of a six-story walkup on Montgomery Avenue, built before city housing legislation prohibited such structures. There the couple raised two daughters, both of whom attended Taft High School, where Helen Schwartz worked as a secretary.

No member of the family was robbed or mugged, and their little apartment was never burglarized. Still, decades of negotiating tight quarters, of trudging up and down flights of stairs, children and groceries in hand, had taken their toll. Nor could anyone ignore how shabby the neighborhood had grown over the years. The forces pushing families from neighborhoods such as Morris Heights to the towers in the east felt powerful and ultimately irresistible. "The whole Grand Concourse just went," Helen Schwartz recalled nearly four decades later. "All the Jewish people were leaving, all my neighbors. I saw them leaving. And I figured, was I any different from them?" In 1970, she and her family joined the ebb tide that seemed to be emptying the entire West Bronx.

To people such as the Schwartzes, Co-op City's attractions were dazzling; the utilities alone seemed reason enough to start packing one's bags. "There was air-conditioning, free electric," Helen Schwartz remembered as she sat in her kitchen nook, not far from a photograph of a serious-looking young brunette wearing a rented bridal gown. For the first time in twenty-five years of marriage, she and her husband did not have to sleep on a couch in the living room. Outside their apartment, the familiar rhythms of the old neighborhood reconfigured themselves in reassuring fashion; people lugged lawn chairs down to the lobby so they could sit and watch the sunset, just the way they used to sit outside their buildings and watch the sunset back home.

So the Schwartzes made the move, along with countless numbers of their friends and neighbors. And thanks to Co-op City's immense scale, changes in the West Bronx occurred more swiftly and traumatically than would have been the case had people trickled away in more leisurely fashion and had so many of them not ended up in a single, highly visible location.

Nevertheless, these families would have gone somewhere. The forces buffeting this population were so deep-seated that huge numbers of West Bronx residents were poised to reroot themselves, whether or not Co-op

City was there to welcome them. Residents of troubled neighborhoods throughout New York City were on the move during those years, headed to Riverdale and Westchester, to Queens and Long Island and New Jersey, destinations that also attracted West Bronx families. Beyond resonating as a symbol, Co-op City simply offered a convenient and tempting destination, and it was easy to follow the people one knew. As Helen Schwartz asked herself so often, was she that different from everyone else?

Many of Co-op City's pioneers, as the first generation of residents were called, came to deeply regret their decision to pull up stakes, as did many others who left the Bronx during those years. It was not that they missed the disarray—by then there was no going back, or any real desire to do so—but that over the years, as the makeup of the complex changed and the pioneers became fewer and fewer in number, they missed one another. As Robert Caro, Robert Moses's biographer, put it, "Lonely was a word you heard all the time."

Columbia University psychiatrist Mindy Thompson Fullilove, whose specialty is the trauma caused by involuntary relocation in cities, uses the term "root shock" to describe the feelings of collective loss suffered by families displaced by urban renewal, and the phrase applies equally well to the emotions of a population whose moorings disappear for other reasons. "It is not just urban renewal that can produce these feelings," Fullilove says. "These were people whose ancestors had experienced two thousand years of Diaspora. What they felt when their neighborhood changed was a profound sense of sadness, a profound sense of dislocation." In situations such as these, Fullilove added, "It's not just a question of winners and losers. In a situation like this, everybody loses."

Compounding the sense of malaise was the fact that, despite the early flush of enthusiasm, Co-op City did not age well. The United Housing Foundation pulled out in the 1970s after a crippling rent strike, and the complex entered its fifth decade mired in physical, financial, and political problems, the unwelcome prospect of privatization hovering like endless bad weather.

For Helen Schwartz, now in her eighties and recently widowed, her daughters long since grown and engrossed in their own lives, the shiny apartment with the parquet floors has become an aerie of regrets. She says she has nothing against her neighbors, a population increasingly dominated by blacks, Latinos, and Asians. Nevertheless, her conversations with them are rare, and she struggles to fill increasingly empty days. It is easy to conclude that the real source of her unhappiness is the isolation and loneliness of advanced old age, aggravated by the death of her husband. Yet she herself is quick to blame her actions of decades earlier. "It's the worst thing I ever did," she said of her decision to join the exodus of West Bronx families to Co-op City. "My friends are all gone. I'd give anything to leave."

⇛ • ⇚

Co-op City forever symbolized the decline of the Grand Concourse. The Cross Bronx Expressway, the six-lane-wide, seven-mile-long concrete and steel monster that smashed its way through the central portion of the borough during the late 1950s and the early 1960s, was blamed for much more.

The road that many people called Heartbreak Highway, first proposed in the 1920s, was envisioned as part of the tapestry of expressways that Robert Moses was weaving across the New York metropolitan area, a critical link between New England and Long Island to the north and east and, via the George Washington Bridge, New Jersey to the west. Drivers hated the Cross Bronx virtually from the day it opened, knowing that it guaranteed a noisy, smelly, and painfully slow journey. The highway did, however, inspire one of the most stirring depictions of human suffering in the twentieth-century city. In a 1974 biography, *The Power Broker: Robert Moses and the Fall of New York*, Robert Caro describes the expressway's impact on the working-class neighborhood of East Tremont, which lay just east of the Grand Concourse, in language that is almost apocalyptic in its furious detail. Shortly before the bulldozers arrived, Caro writes, a group of East Tremont residents, seeking a glimpse of what their future held, visited an area where construction had already begun. There they stumbled into a nightmare.

> Where once apartment buildings or private homes had stood were now hills of rubble, decorated with ripped-open bags of rotting garbage that had been flung atop them. . . . Giant wreckers' balls thudded into walls; mammoth cranes snarled and grumbled over the ruins, picking out their insides. Huge bulldozers and earth-moving machines rumbled over the rubble; a small army of grime-covered demolition workers pounded and pried and shoveled. . . . And in the midst of this landscape of destruction, a handful of apartment houses still stood. . . . Going inside, they found the lobbies littered with shards of glass that had once been big ornamental mirrors and with the stuffings

from the armchairs and sofas that once had been their decoration, and smeared with excrement not only animal but human, from winos and junkies who slept in them at night. Stumbling upstairs, the committee found the doors to many apartments ajar; through them, they could see empty rooms, walls ripped open by vandals who had torn the plumbing pipes out of them. Other doors, however, were closed and locked. . . . And behind those doors the committee found people, not winos but respectable Irish or Jewish families like themselves.

Far worse lay ahead. Within five years after completion of the portion of the expressway that ran through East Tremont, the neighborhood was a human graveyard.

Windows, glassless except for the jagged edges around their frames, stared out on the street like sightless eyes. The entrances to those buildings were carpeted with shards of glass from what had been the doors to their lobbies. . . . Plaster from the walls lay in heaps in corners. . . . Staircases were broken and shattered. Banisters had been ripped from their sockets, for scrap and a fix if they were iron, for malice, an expression of hatred and revenge on an uncaring world, if they were wood. Raw garbage spilled out of broken bags across the floor. . . . There was no heat in those buildings; if they were homes, they were homes as the cave of the savage was a home. And yet these *were* homes—homes for tens of thousands of people. They were homes for welfare tenants and for the poorest of the working poor, . . . for mothers who say desperately to the stranger, . . . "I got to get my kids out of here."

One evening a few years ago, Caro sat in an armchair at the Century Club, just off Fifth Avenue, and talked about his visits to those buildings. "Interviewing those families was the worst experience of my life," he recalled. "The drug dealers, the gaping holes in the buildings, the disgusting stench of urine. The elevators never worked. You heard strange noises. If I stayed after it got dark, I was truly scared. The whole experience was unrelievedly scary and horrible." He used to park on the street, and on every visit, he prayed he would make it safely back to his car.

More important, Caro added, were he writing the book today, he would again come to the conclusion that the mammoth highway represented the single most powerful force leading to the collapse of the southern half of the borough, especially the breakdown of communities such as East Tremont.

"It's a cliché to say the neighborhood would have changed without the expressway," Caro said that evening in the Century Club. "The neighborhood didn't have to become a slum." He pointed out that Belmont, the Italian American enclave north of East Tremont, has endured; had East Tremont been left alone, he contends, it too would have survived. "People say that what happened to East Tremont was inevitable. I don't agree."

Caro is an immensely appealing person and a persuasive talker, especially when it comes to the role of the highway that without question left a profound mark on the borough. And given the magnitude of his achievement, it is tempting to accept his interpretation with regard to Moses and the city he shaped so profoundly. But as an understanding of the forces that bedeviled the Bronx has evolved, to place heavy blame on the Cross Bronx Expressway for the borough's misfortunes, at least to the extent that Caro does, is increasingly difficult.

With ever greater vigor, social scientists have questioned Caro's methodology, his statistics, his sources, and his reading of history. In *Robert Moses and the Modern City: The Transformation of New York*, an encyclopedic work published in 2007 that brings together much of the revisionist scholarship on Moses, Ray Bromley, a professor of geography and planning at the State University of New York at Albany, argues convincingly that the expressway would probably have been built with or without Moses and was largely inevitable, given the placement of the George Washington Bridge.

More centrally, Bromley points out that some of the Bronx neighborhoods hardest hit during the 1970s, such as Melrose and Morrisania, were not near the expressway and that communities in Brooklyn such as Brownsville and East New York, untouched by superhighways, were also ravaged during those years. Although the Cross Bronx admittedly left disruption and devastation along its route, he notes that only the rare large-scale public work arrives without a human cost, one increasingly forgotten over time. "Clearly something much broader than the negative impact of expressways on neighborhoods was at work," Bromley concludes, "the results of decades of disinvestment and suburbanization compounded by urban renewal, deindustrialization, racism, fraud, and city, state, and federal policies favoring Manhattan and the suburbs."

As part of the sweeping reassessment of Moses's legacy, Columbia University historians Hilary Ballon and Kenneth T. Jackson, the editors of *Robert Moses and the Modern City*, argue that Moses's embracing networks of highways, bridges, and other ambitious public projects were precisely the building blocks that helped make possible the modern metropolis and that the economic and psychological turnaround that buoyed New York City in the 1990s could not have occurred without his outsized creations. Despite the influence of urbanologist Jane Jacobs and her advocacy of a city composed of small, finely textured communities, some historians have suggested that New York could use a few master builders today; in particular, they add, those charged with redeveloping Ground Zero would do well to emulate Moses's vigorous and single-minded pursuit of his goals.

Even Moses's most passionate defenders agree that the Cross Bronx Expressway, like Co-op City, accelerated change and made that change infinitely more traumatic. Yet what if the huge highway had never been built? How then would the West Bronx have fared?

Marshall Berman, the political scientist who grew up a few blocks east of the Grand Concourse, understands the temptation to blame the great gash that is almost literally in his old backyard for the upheaval of the streets of his youth. And like Caro, he describes the road's impact with eloquence and passion. "For ten years," Berman writes in *All That Is Solid Melts into Air*, "through the late 1950s and early 1960s, the center of the Bronx was pounded and blasted and smashed. My friends and I would stand on the parapet of the Grand Concourse, where 174th Street had been . . . and marvel to see our ordinary nice neighborhood transformed into sublime, spectacular ruins."

The apartment houses directly on the Grand Concourse were spared by the highway, which passed beneath the boulevard. Yet Berman, in his mind's eye, still sees the image of "hundreds of boarded-up abandoned buildings and charred and burnt-out hulks of buildings; dozens of blocks covered with nothing at all but shattered brick and waste." The decals of curtains and window boxes and flower pots that the city posted on the windows of these sealed-up apartment houses seemed the final indignity.

And yet, like Moses's more recent defenders, Berman wonders, how would the borough have fared had there been no road? "How many of us would still be in the Bronx today, caring for it and fighting for it as our own?" he asks. "Some of us, no doubt, but I suspect not so many, and in any case—it hurts to say it—not me. For the Bronx of my youth was possessed, inspired, by the great modern dream of mobility. To live well meant to move up socially, and this in turn meant to move out physically; to live one's life close to home was not to be alive at all." Reluctantly, Berman concludes, "All through the decades of the postwar boom, the desperate energy of this vision, the frenzied economic and psychic pressure to move up and out, was breaking down hundreds of neighborhoods like the Bronx, even where there was no Moses to lead the exodus and no Expressway to make it fast."

$$\Rrightarrow \bullet \Lleftarrow$$

For decades, historians have debated the impact of Co-op City and the Cross Bronx Expressway on the West Bronx, and they will probably do so for decades to come. In the opinion of many families who lived on those beleaguered streets, however, the drumbeat of newspaper articles describing quickening racial, ethnic, and socioeconomic changes and depicting the rapidly deteriorating face of a beloved neighborhood did almost as much to hasten the decline as changes born of bricks and mortar. So powerful was the impact of some of these articles that local residents quoted them verbatim decades after they were published, excoriating them over and over for having helped kill a community.

As late as February 1965, the *New York Times* published an article about the Grand Concourse under the moody but unthreatening headline "The Broadest Boulevard of the Bronx Evokes Memories and Sometimes a Bit of a Sigh." Within seventeen months, the tone of the paper's coverage had changed dramatically. An article that appeared on July 21, 1966, bearing the headline "Grand Concourse: Hub of Bronx Is Undergoing Ethnic Changes," painted a bleak portrait of a once-thriving world.

"The neighborhood is deteriorating, there's no getting away from it," the reporter, Steven V. Roberts, quoted a local rabbi as saying. "We had 140 families seven years ago, and now we have 60. Even the big synagogues on the avenue are having trouble." A kosher butcher added, "Three shops closed around here last month. You couldn't buy a store on this block 10 years ago for any amount of money. Now they'd give it to you."

Roberts cited statistics predicting that the neighborhood, which had been 98 percent white in 1950, would be less than 50 percent white by 1975. "City officials fear that the area is on the edge of panic," he wrote. Those unnamed officials "emphasize that they do not share the idea that Negroes inevitably cause a neighborhood to deteriorate. However, they say, the infusion of Negroes often produces reactions in a community that can lead to rapid decline."

Roberts took pains to point out that despite considerable local antipathy to the black and Puerto Rican newcomers, the real engine of change involved the rejection by young white families of the world of their parents. Nor was Roberts the first to parse the racial underpinnings of tensions in the West Bronx. "Let's face it," one public school teacher was quoted as saying, "for most of these parents, a good school is a white school." And a longtime resident added, "If Negroes and Puerto Ricans come in, there is no incentive to keep up the buildings. Maybe that's prejudice, maybe that's bigotry, but that's what happens."

To many readers of the *Times*, who were growing accustomed to accounts of an increasingly troubled city, the article would not have struck a particularly personal chord. But in the West Bronx the piece landed like a bombshell. Roberts's account of the changes roiling the area provoked fury, panic, and most of all the sinking feeling that the reporter had painted an accurate portrait of the community. For so prestigious a publication to conclude so flatly that the West Bronx was in serious distress and that issues involving race lay at the heart of the problems gave shape and reality to what people had long suspected but hardly dared let themselves admit. The newspaper held a mirror up to the community, and the community was terrified by what it saw.

Roberts, long departed from the *Times*, says he has no real memory of either his article or the storm it provoked. Still, to many people in the West Bronx, his words were as chilling as Co-op City's towers or Moses's rumbling expressway, in that they confirmed what large numbers of residents already knew. If people needed an excuse to flee the West Bronx, articles like this provided one.

$$\equiv \quad \bullet \quad \equiv$$

Many of those who left, such as the family of Sam Goodman, had roots on the Grand Concourse that went back almost as far as the boulevard itself. Goodman's grandparents on both sides were among the street's earliest residents, having arrived shortly after the Jerome Avenue subway line opened in 1918. In 1927, his maternal grandparents moved into the grandly named Rockwood Hall at 1555 Grand Concourse, near Mount Eden Avenue, to a sprawling, seven-room apartment so spacious that the dining room table could accommodate ten grandchildren and all their parents. The two-story lobby, with its marble mantelpiece, brass elevator doors, thick Oriental rugs, and coffered ceiling embossed with dark wood octagons, had the aura of Camelot about it. Goodman's mother grew up in Rockwood Hall, and when her son was born in 1952, virtually all his relatives lived within walking distance of the boulevard.

A small black-and-white snapshot taken in 1955 shows a solemn-eyed child wearing a dressy coat and hat and standing in front of the Lorelei fountain. Seven years later, that same child, now struggling to master a three-speed English racer, discovered another face of the Grand Concourse. Every morning for months, he woke up early and pedaled along the boulevard toward Mount Eden Avenue, with its bakeries, candy stores, and fruit stands. He remembers the giddy sense of being awake when most of the neighborhood was asleep, the sound of traffic barely a whisper. A nickel bought him a warm onion bagel with cream cheese cut from a large block in the display case. Then he pedaled home.

The family was vacationing in Washington, D.C., when the article by Steven Roberts appeared in the *Times*. Goodman's parents read the words about rundown buildings and empty synagogues and stared at the photograph of the vacant kosher butcher shop on 161st Street. "The *Times* article scared the bejesus out of the neighborhood," Goodman remembered, and it so terrified his mother that she immediately contacted her congressman, James Scheuer, to beg him to tell her frankly whether the newspaper's dire predictions were accurate. Sitting with Scheuer in the congressional dining room, she got her answer. "In twenty years," the congressman told his panicky constituent, "we're going to bulldoze the whole place down." By October, the Goodman family had moved to Connecticut.

Some three decades after those magical early-morning bicycle rides, Sam Goodman returned to the borough of his birth. He lives in a co-op at 800 Grand Concourse, opposite the courthouse, and works as an urban planner in the borough president's office. In describing why a neighborhood seemingly so solid proved so vulnerable to convulsive change, he can reel off the usual suspects: the disinvestment, the public and private abandonment, city and federal policies that hammered vulnerable communities, a dysfunctional welfare system, the scourge of drugs, a lack of jobs, racial prejudice, greedy landlords, and even greedier insurance companies. Nevertheless, Goodman understands better than most people the forces that gave a speed and an inevitability to the transformation of the West Bronx.

"The neighborhood grew up so fast, and it died so fast," he said. "It's like a car. The transmission goes, and suddenly everything goes. It's the same thing with a community. All the people and all the buildings were of the same generation, and so everything wore out at the same time." That the newcomers were of a different class and, in a still largely segregated city, of a different color only exacerbated the problem. "Because the neighborhood was so homogeneous," Goodman said, "anyone who didn't fit was an outsider."

"Bends in the Road"

ONE UNSEASONABLY WARM EVENING in the autumn of 2005, a group of Bronx residents came together to talk about the place where they lived. The gathering, held at the Bronx Museum of the Arts, was the first in a series of events examining how a thoroughfare draped in memories could take its place in a twenty-first-century metropolis so different from the one in which it had come of age. Although the centennial of the Grand Concourse lay four years in the future, the events seemed to sum up all the issues facing the boulevard as the street and its surrounding neighborhoods grappled with their complicated past and looked ahead to their future.

Judging by the events over those four days, the future of the West Bronx seemed wreathed in questions. How could the area be made welcoming to a new population that was so different from the original settlers but was, in many instances, arriving with similar dreams and aspirations? How could such a transformation take place while still acknowledging and respecting the heritage and contributions of those who came before? And perhaps most challenging of all, how could an area so battered for so many years continue its trajectory toward health and renewal?

The answers were complicated and often elusive. Despite the continuing signs of renewed health, not every indicator was rosy. The story of the Grand Concourse in the early twenty-first century is a story of successes interlaced with problems. Although there are signs of renewal amid the decay, development is uneven. Although the boulevard is clearly enjoying a revival, its long-term future is by no means guaranteed. Many of the elements that helped the Grand Concourse grow and flourish at the beginning of the twentieth century are present today, but the extent to which they will be supported and nourished remains an open question, especially in the face of the cataclysmic economic turmoil that was buffeting the city, the nation, and the globe as of late 2008 and early 2009.

Nevertheless, if one were to predict the area's future, the prognosis would be far more positive and encouraging than at any point in the past half century. In a reflection of that optimism, the program at the museum opened on a high note with a presentation by Bronx Borough President Adolfo Carrión Jr., who had made reclaiming the borough's most visible street a signature effort of his administration. Looking dapper in a dark suit, Carrión escorted the audience through high points of the boulevard's history, among them Governor Al Smith's optimistic remarks at the opening of the Concourse Plaza Hotel and decades' worth of Yankee victories at the ballpark down the hill. Then he described the city's ambitious plans to recapture some of the charm and graceful proportions of Risse's late-nineteenth-century design of the street, through the addition of trees, flowers, benches, bike lanes, cobblestone sidewalks, widened medians, and decorative lighting. The goal was to refurbish a ten-block stretch north of the courthouse in time for the street's one hundredth birthday, to be marked officially on November 24, 2009. Earlier that year, the Bronx would also celebrate the opening of a new $1.3 billion Yankee Stadium just steps from where the powerhouse slugger from Baltimore hit his first home run wearing pinstripes on a blustery April day in 1923.

Events such as the gathering in 2005 invariably tap into a vein of almost suffocating nostalgia. The yearning to return, at least in memory, to the era in which stars twinkled at the Paradise and wedding guests danced the fox trot at the Concourse Plaza Hotel, the moment that Jews wearing exquisitely tailored suits paraded down the Grand Concourse and lingered in Joyce Kilmer Park, reluctant for the day to end, casts a powerful shadow on the present-day boulevard.

Mental snapshots of that era bring unexpected comfort. On a Web site devoted to Bronx nostalgia, a woman told of a friend who used to lull herself to sleep by traveling in her mind back to her old apartment at 2024 Morris Avenue, a few blocks off the Grand Concourse. "In her mind, she walks up the flights of stairs, noting all the apartments on each landing, and recalling who lived in them," the friend remembers. "Then she finally makes it to her apartment on the top floor, and walks through the rooms as if it was yesterday." That so many buildings that shaped this environment remain standing gives special poignancy to this lost world and represents a teasing, taunting reminder of what once was.

Some of these emotions were on display that evening at the museum; a longing partly for one's past and especially for one's youth permeates the boulevard's very bones. Yet for the first time in decades, another generation was chiming in to extol the street's charms. One of the most ardent cheerleaders for the new face of the West Bronx is a balding man in his late thirties named Michael Bongiovi, and that evening at the museum, as guests milled about the lobby drinking wine in paper cups, he talked about why he, a boy from Brooklyn, had made the move to the borough's most celebrated thoroughfare. He and his partner, Bill Madden, both of whom held administrative jobs at Fordham University, had closed the previous winter on a two-bedroom co-op on the twenty-third floor of Executive Towers, and by the following spring they were admiring the tulips in Joyce Kilmer Park, just visible from their terrace. Judging by how quickly prices in the building were escalating, they suspected that their apartment was already worth much more than the $135,000 they had paid for it.

Bongiovi chronicled his journey to Executive Towers on one of the many Web sites devoted to Bronx nostalgia, and in response, expats with names like Margie and Marilyn and Trudy sympathized with his search for a decent local supermarket and his struggles with the co-op board ("I feel like it's 1949!" Bongiovi complained at one point). Some members of this Diaspora even expressed a longing to join him. "I envy you in a way," a woman named Laura wrote in a posting. "My days in the Bronx are a distant memory now. . . . If it weren't for having a family, tho, I think, if it was just me, I'd be back in a flash."

Bongiovi and his partner have emerged as poster children for the boulevard's changing face, examples of middle-class whites who could live anywhere in the city but choose to make their home here. They are fixtures at events such as the one at the museum, and they are outspoken about their desire to sink roots in the borough. "This apartment is plenty big enough for a child," Bongiovi pointed out as he sipped his wine, and he insists that he will remain on the Grand Concourse until he dies or retires to Florida, whichever comes first.

⟹ • ⟸

Bongiovi and his partner are not the only people boasting of an address on the Grand Concourse in recent years. Even some who arrived during the street's most difficult period and traveled very different journeys succumbed to its spell.

When eighteen-year-old Millie Lopez caught her first glimpse of the Grand Concourse on the day in 1972 that she visited an office on Walton Avenue, a few blocks west of the boulevard, to get a Social Security card, she felt like the proverbial Dorothy arriving at the Land of Oz. "All those pretty buildings, all those white people—it was like something foreign," she told a reporter. "I thought I was in another city."

Lopez was the youngest of six children—her family being that amalgam of Puerto Rican and New Yorker that calls itself New Yoriquen. Growing up in a tenement on 138th Street, the only world she knew was the South Bronx that had begun unraveling in the 1950s, a world of rooftop rapes and sirens in the night. For her, a move to the Grand Concourse was a sign of upward mobility, the logical next step after the acquisition of a college degree, a car, and a job with the Community Service Society, one of the city's major nonprofit agencies. Like generations before her, Lopez proceeded to the boulevard in stages, first to the projects, then to Walton Avenue, and by 1980, to a sunny apartment at 1001 Grand Concourse, overlooking Yankee Stadium. Just writing the address gave her a slight thrill.

Although Lopez did not have a sunken living room, she did have a doorman, along with parquet floors so bright she could practically see her reflection in them. Like Bongiovi more than two decades later and like generations of boulevard residents before her, she was enchanted with the view. "I feel grateful to live on the Concourse," Lopez said as she gazed out her window at the diamond-bright vista in the distance. "I feel very special, very privileged." The crime that terrified many of her elderly neighbors was nothing compared to the mayhem she had known. "For the first time in my life, I come home late at night and I don't worry," she said. "Compared to the South Bronx, this is heaven." Her euphoria at times echoed almost word for word that expressed by many of the borough's earlier residents and underscored the fact that despite the pounding the boulevard suffered over the years, the Grand Concourse never ceased exerting a peculiar power over many who lived along it.

⇒ • ⇐

Despite such enthusiasm, the neighborhood is still struggling to shed the vestiges of its nightmare past, and not all strolls down the boulevard's memory lane these days are drenched in warmth. The evening after Carrión's upbeat presentation at the museum, at a panel discussion entitled "Bends in the Road: Looking Forward and Backward along New York's Grand Concourse," conversation shifted to what has become an increasingly divisive local issue, the conversion of ground-floor spaces in residential buildings on the boulevard into commercial establishments. Starting in the 1980s, a rainbow assortment of signs and awnings began sprouting along the southern half of the Grand Concourse, announcing the presence of Ecua Musica, Mariama Hair Braiding, Concourse Jewelry Exchange,

Pay-o-Matic Check Cashing, and countless other small enterprises. Immigrants from Ghana have planted an especially noticeable flag: numerous storefronts accommodate businesses designed to help them buy houses in their homeland, a symbol of prestige for this West African community.

Historians, preservationists, and others eager to recapture the street's midcentury glory see these outcroppings as blights on the landscape, a dispiriting reminder of how completely the street and its adjacent neighborhoods have been transformed in recent decades. Others more sympathetic with the largely minority entrepreneurs who run all these bodegas and botanicas and live-poultry markets contend that such uses represent signs of commercial vitality. More to the point, they add, for a mother dashing out late at night to buy a carton of milk, a bodega around the corner is extremely convenient, even if the sign isn't very attractive and even if by law the bodega shouldn't be there in the first place.

$$\Longrightarrow \bullet \Longleftarrow$$

Conversations about rifts that continue to define life along the Grand Concourse grew even edgier during a discussion of how issues of race and housing have shaped the boulevard's fortunes. As Brian Purnell, research director of the Bronx African-American History Project at Fordham University, a pioneer effort that is giving voice to people whose stories have traditionally been excluded from the borough's history, reminded the audience pointedly, "The Grand Concourse wasn't grand for everyone." For the African American women who offered their services at the Bronx Slave Market, for the boys shooed off the boulevard by armed policemen, for young black couples refused apartments on the Grand Concourse, and especially for the countless African American and Puerto Rican families caught up in the turmoil of the 1960s, 1970s, and early 1980s, the name Grand Concourse seems almost ironic. "I don't care about the Grand Concourse," one older African American man volunteered during that session. "My parents couldn't get an apartment there, and I never lived there." He was thirteen the first time cops stopped him on the boulevard, he said, and, as he remembered the encounter, they told him what would happen if he lingered: "We'll break your black ass. You'll never walk again."

As more is learned about the forces at work in the Bronx in recent decades, and as an awareness of who suffered and how much they suffered has deepened, feelings of outrage and resentment have grown also. Was local antipathy by whites toward blacks worse than the sometimes violent skirmishes between Jews and Irish Catholics that defined so many Bronx boyhoods? Would the street have evolved differently if earlier on it had been more welcoming to minorities, many of whom lived so close and would have given anything to occupy its apartments or even stroll undisturbed down its wide and tranquil sidewalks? Would the presence of a more mixed population have cushioned the blows inflicted on this community? Like many questions about the boulevard's complicated history, these lingered uneasily in the air.

Yet whatever the answers, no one could ignore the fact that by the autumn of 2005, the Grand Concourse was enjoying a vitality not seen in half a century. The economic resurgence that had produced a demand for housing outside Manhattan—first in Brooklyn, then in Queens— was reaching into the southern tip of the Bronx, pushing north along the Grand Concourse and edging onto its side streets. Shifting patterns of crime and drug use, combined with sweeping demographic changes, had been augmented by smarter and more sophisticated policing and the efforts of two energetic and politically ambitious borough presidents. Although media attention focused on the renaissance of Charlotte Street, the onetime heart of devastation north of Manhattan, the West Bronx had long since begun witnessing its own rebirth and reemerging as a potent symbol of the borough's identity.

Much of the change came through the work of community development corporations that tapped sources of federal and other money to rehabilitate rundown buildings and usher them into happier and healthier existences. Notable among these groups was the Northwest Bronx Community and Clergy Coalition, an umbrella organization that battled banks, insurance companies, and officials on all levels of government and whose arsenal included a highly unusual landlord: a short, middle-aged Polish Jew named Joseph Bodak. In a 1988 photograph in the *New York Times,* Bodak is seen bundled in an overcoat, gloves, and a dark fur hat, standing in front of Thomas Garden, one of dozens of local buildings he managed. Over the years, Bodak renovated and ran more than forty apartment houses south of Fordham Road, to almost universal acclaim. As one of his tenants told a *Daily News* reporter in 1982, "With Mr. Bodak, you say there's a leak in the ceiling, you see the plumber the next day."

The 1980s, the era that saw the Grand Concourse begin its long march toward better days, witnessed other significant events. The year 1988 marked the official end of abandonment along the boulevard; this was the year private developers began restoring the last five vacant buildings on the Grand Concourse, situated on a five-block stretch south of 170th Street.

By the start of the twenty-first century, a growing number of star architects were leaving their imprint on and near the boulevard. Rafael Vinoly, creator of an eye-catching housing court on the Grand Concourse at 166th Street, was putting the finishing touches on his Bronx Hall of Justice, a $325 million, two-block-long, glass and aluminum criminal courthouse scheduled to open in two years' time on 161st Street just east of the boulevard. The Bronx Museum of the Arts was awaiting a $19 million expansion by the high-profile Miami design firm Arquitectonica; the museum, once the home of Young Israel of the Concourse, one of the area's beloved synagogues, would be sheathed in a pleated aluminum façade and its interior space nearly doubled. From Hostos Community College, a branch of the City University system on the lower Grand Concourse, up to Bronx-Lebanon Hospital Center at Mount Eden Avenue, institutions all along the boulevard were being expanded, improved, and shined up—often literally, as in the case of the old courthouse.

On July 8, 1999, after a ten-year battle and a century to the day after its original dedication, the Lorelei fountain was rededicated and moved from the northern part of Joyce Kilmer Park to a site on the park's southern end. With the help of $310,000 in city and foundation money, artisans had crafted new heads and fins for the mermaids, and nearly one million dollars in municipal funds made possible the monument's complete overhaul. For the first time in decades, water splashed over the mermaids' sleek marble bodies.

The building that was once home to Temple Adath Israel, the date inscribed on the cornerstone a reminder of its original incarnation, pulsed with life as a Seventh-Day Adventist temple, poised to celebrate its thirty-fifth anniversary two years hence. On Friday nights, worshippers climbed the marble steps, pushed open the great brass doors, passed beneath glowing images of the burning bush and the lion of Judah—Old Testament icons that had a place in their own faith—and prayed beneath a ceiling of blue and gold medallions, just as well-heeled Conservative Jews of the West Bronx had done for half a century.

Almost uniformly, the boulevard's showplaces were in vastly better shape than they had been three decades earlier, thanks partly to federal subsidies for low-income families that made the phrase Section 8 housing one of the most familiar terms in the political lexicon. By 2007, an article in the New York Times touted a one-bedroom apartment on the Grand Concourse in Highbridge for $90,000. The following year, the newspaper described a three-bedroom apartment on the boulevard in the Bedford Park section that had sold for $310,000.

By far the most dramatic turnaround had taken place at Roosevelt Gardens, long a symbol of the area's grimmest days. After sitting vacant for three years, the building was purchased by the Kraus Organization, which gutted and rehabilitated the apartments and reopened them in 1980 as subsidized units. So attractive was the newly refurbished complex that when Kraus began renting the 291 units, the company received ten thousand applications.

Yet at Roosevelt Gardens as at virtually every location along the boulevard, reminders that the street had a long way to go were impossible to ignore. The array of video surveillance screens and the security guard in the prisonlike kiosk in the courtyard attested to continuing problems. And whether Roosevelt Gardens can ever entirely shake its calamitous history is uncertain; many people familiar with that history find it hard to pass the complex without being reminded, at least momentarily, of eight-year-old Missy Holden, literally beheaded after stepping into an elevator on the way to her apartment.

The Concourse Plaza Hotel reopened in 1983 as subsidized housing for the elderly under the sponsorship of the Mid-Bronx Senior Citizens Council, one of the community groups that have done heroic work to reclaim Bronx neighborhoods in distress. But the Andrew Freedman Home, run by the same organization, remained empty except for the social services groups encamped in the basement, awaiting its next act.

Noonan Plaza, the palatial apartment house in Highbridge that hit bottom in 1980, had been rescued in 1982 by the Glick Construction Company and renovated to provide nearly three hundred subsidized apartments. Ironically, the once magnificent courtyard played a major role in updated security arrangements. "It's designed like a fort," Frederick Ginsbern, the son of the original architect, who helped with the renovation, told a reporter. "It's great for security now."

The Lewis Morris, the best address in the West Bronx for half a century, was again filled with families. Nevertheless, the ghostly footprint of a long-disappeared fireplace on the inlaid marble floor in the lobby offered mute evidence of the décor that had been lost, and a security cage encased in bulletproof glass spoke to the fact that the building was hardly in great shape. Outside the entrance, broken glass crunched underfoot, and on gusty days, the wind whipped about wrappers from Wendy's, Popeye's, McDonald's, White Castle, KFC—seemingly every fast-food place in the city.

Up and down the boulevard, roll-down metal store shutters, boarded-up windows, and courtyards transformed into dust bowls testified to the area's continuing needs. Passengers riding the bus along the Grand Concourse were treated to a recorded announcement urging them to carry valuables in an inside pocket. The stench of urine permeated subway stations all along the thoroughfare. So sketchy was the ambiance on some blocks that crews from the television series *Law & Order* still trooped up to the Grand Concourse to film episodes set in troubled neighborhoods.

⇒ • ⇐

To people familiar with the boulevard's storied past, the pace of residential revival along the Grand Concourse might have seemed spotty and in some cases deeply frustrating. But if ever an undertaking suggested a bright future for the West Bronx and offered a template for how the area's past and present strengths could be merged, it was the rebirth of the theater that many people regarded as something of a shrine.

The reclamation of Loew's Paradise is one of the undisputed success stories of the Grand Concourse, and the highlight of the series of events that began with the borough president's speech at the Bronx Museum of the Arts was a reception at the almost completely restored playhouse. After countless false starts and broken promises, the Paradise was poised for its transformation from a Depression-era movie palace to a venue for Latin-music concerts and live boxing matches. In 1999, a Westchester County developer named Richard DeCesare rented the building and began pouring millions of dollars into recapturing the theater's original luster—purchasing four thousand burgundy velour seats, scrubbing the caryatids and cherubs, and repainting the midnight-blue ceiling. Financial problems prevented DeCesare from renewing his lease, but in 2003 a more fortunate developer named Gerald Lieblich took over the Paradise, and on the evening of the reception he mingled with the crowd and proudly accepted congratulations for what by many measures was a significant step forward in the West Bronx.

With finishing touches still to come in preparation for an official opening the following month, visitors stepped gingerly around paint-splattered drop cloths and puddles of dirty water that had collected on the great stage. Huge sheets of plaster hid the gold velvet curtain, and dust and shadows swallowed up much of the vast space. Yet the mood was euphoric. Men and women, some of whom remembered the theater from their childhood, others of whom were barely alive when the Paradise presented its last picture show, wandered about as if in a trance, necks craned and eyes turned upward, sighing to one another and sometimes to themselves.

No goldfish were in evidence, nor had clouds returned to the heavens. But in virtually all other respects, the architects had re-created the building's original opulence. Glowing chandeliers and lighting fixtures, some shaped like giant dragonflies, bathed the onetime Italian Baroque garden in golden light. Marble balconies and carved ceilings, every speck of grime wiped away, looked as pristine as they had in John Eberson's day. In the mezzanine, visitors ran their fingers along the bright brass railings. "It looks like the Venetian Room in Las Vegas," one man volunteered. In the shadowy auditorium, a pair of women gazed up at the ceiling, trying to find the stars.

⇛ • ⇚

Sam Goodman, the urban planner whose life has been entwined with the area going back several generations, and who was a major participant in the series of events sponsored by the Bronx Museum of the Arts, has been watching the borough's shifting fortunes for half a century. One day not long ago, Goodman sat in a coffee shop on 161st Street just off the Grand Concourse, among the few places near the ballpark where a waitress will serve a sit-down meal, and talked about the future of the West Bronx. Cranes preparing the site for the new Yankee Stadium moved languidly in the distance—dark, elongated fingers tracing the borough's future on the late-afternoon sky.

These days, Goodman sees much cause for optimism. The fate of the West Bronx and by extension the Grand Concourse is intricately linked to that of the Bronx as a whole, he pointed out, and with its strategic location and infrastructure, its situation could not be more advantageous. The Bronx sits at the nexus of a regional transportation system in which an extensive network of highways, subways, and commuter rail lines is already in place. As people continue to move farther from the city's heart, he sees these attributes as increasingly critical.

Of equal importance, in Goodman's view, is an unparalleled housing stock—the huge number of large apartments in still-sturdy buildings that, paradoxically, were protected from the wrecker's ball by four decades of disinvestment. "An eighteen-hundred-square-foot apartment in a six-story building—you can't build that today," Goodman noted. Evolving tastes are buttressing economic forces; the growing affection for Art Deco is producing a renewed appreciation of the aesthetics of these apartment houses.

In a city of ever greater racial and ethnic variety, the borough's diversity is seen as another source of strength, at least in the long run. More than half the residents of the Bronx are Latino, and with each passing day the mix is thickened by immigrants from around the globe, not just Latin America but also Asia, Africa, Eastern Europe, the Middle East, and the Caribbean.

These shifts are occurring at a moment when race and ethnicity, such defining factors until well into the 1960s, are less of an issue when it comes to neighborhood stability. New Yorkers are increasingly comfortable in a multicultural metropolis, and in a city less defined by race and ethnicity, change in a community's makeup does not automatically undermine its social fabric—sometimes just the opposite is the case. That the Bronx remains one of the few places where New Yorkers earning a living wage can afford to invest in housing whose value is almost certain to increase, at least in the long run, only enhances the borough's appeal.

These assets are considerable. And if one mantra is repeated over and over by public officials, urban analysts, local residents, and people who simply wish the best for the Bronx, it has to do with the borough's "rebirth." Each new café, each co-op conversion, each gallery opening is heralded as a sign of renaissance.

Often overlooked in the euphoria is how much the borough continues to struggle economically, and not just because of the economic problems that were washing over the city as of early 2009. According to the U.S. Census, as of 2006, nearly a third of the Bronx's 1.3 million residents live below the federal poverty line, a level so modest that a family of four was considered officially poor only if its annual income was under $19,300. Even more telling is the fact that of residents under age eighteen, 41 percent live in poverty. By every measure of hardship— unemployment, infant mortality, educational level, school drop-out rates—the borough lags behind the rest of New York City and much of the nation; the Bronx has the sad distinction of being the poorest urban county in the United States. Though much of the poverty is concentrated in neighborhoods that constitute the traditional South Bronx, the impact radiates throughout the borough. And although Fordham Road is the busiest commercial strip in the borough, home to a number of national chains, many businesses are reluctant to set up shop in the Bronx, particularly in the blocks south of Fordham Road. Especially in the West Bronx, the presence of all these needy families could ultimately prove a powerful economic drag.

"The closest place to get a fresh bagel is Lexington Avenue and Eighty-sixth Street," Goodman pointed out that day in the coffee shop.

> South of Fordham Road, there's no bakery where you can buy fresh bread. After eight p.m., only three places to eat are open on 161st Street near the stadium, and they all have grates and bullet-proof glass. There are only three bookstores in the borough, and none of them are in this neighborhood. There's no gym, no sports club. And there are so few bank branches that if you need to see a teller on a Friday afternoon, you have to stand on line for forty-five minutes.

Many of the Bronx's economic woes translate into housing problems that rival those of the borough's worst years. In 2007, even as the city's economy was enjoying its happiest moments, a five-story walkup on Morris Avenue and 169th Street, just three blocks east of the Grand Concourse, was without reliable heat and hot water for months, according to an article in the *New York Times*. The reporter described conditions under which the two dozen tenants were living as "a dismal, surreal housing arrangement that seems as much out of Kafka as Dickens" and noted that the structure had recently been added to the list of the city's two hundred worst maintained buildings, cited for 181 of the most serious housing code violations.

Young families eager to raise children in the area face especially daunting obstacles. The necklace of high-performing public schools that educated generations of West Bronx children collapsed in the 1960s, and Community School District Nine, which covers the heart of the neighborhood, has long been one of the city's most troubled districts—and judging by articles about its activities in the *New York Times* and other publications, it has also been among the most corrupt.

⇒ • ⇐

Yet despite all the negative numbers and what everyone agrees is a long and arduous journey ahead, particularly given the economic upheavals that began in 2008, conversations about the Grand Concourse these days tend to conclude on a mostly upbeat note, and rightly so.

"When I see the construction of Yankee Stadium and think about everything else that's happening on the Grand Concourse and 161st Street," said Wilhelm Ronda, the chief of urban planning in the borough president's office, "I get the same feeling people must have had in 1923, when the stadium and the Concourse Plaza Hotel were built. I walk along 161st Street, and I relive that year. I feel so fortunate to be here at this moment, when there's so much investment and so much is happening."

One person who would have shared Ronda's optimism was Donald Sullivan, director of the graduate program of urban planning at Hunter College. Sullivan, who grew up in what was known as the traditional South Bronx, died in 1989 at the age of forty-seven. He did not make it to an address on the Grand Concourse. Nevertheless, during his relatively brief lifetime he had a powerful sense of the neighborhood's potential, and more than any other single event, the 1976 exhibition on the Art Deco legacy of the West Bronx that he and Brian Danforth curated at Hunter College helped draw attention to the glories of these apartment houses by articulating in compelling fashion the threats they faced and sounding a clarion call for their rescue.

A decade before his death, Sullivan spoke to a reporter about the future of the Grand Concourse and its neighborhoods. "This could be one of the most glorious interracial low-income and middle-income communities," Sullivan said in that conversation. "It has the architecture, the topography, the parks, the river, direct train service, beautiful large apartments." In his mind's eye, he envisioned a renewal of the West Bronx, a revitalized Grand Concourse at its heart; he saw the Art Deco buildings, solid as in their prime, restored to their old polish, the streets sparkling, tulips blooming once more in Joyce Kilmer Park. The Concourse Plaza Hotel, he pointed out, was still just minutes from Times Square. "There are three hundred thousand people living in the West Bronx," Sullivan said, "and for them the Concourse has always been the focus. They were all fine streets, but this was *the* address, and still is. For all of us, the dream continues."

≡ • ≡

As the Grand Concourse and the West Bronx look toward their future, no one underestimates the tasks going forward. There is an understanding that bricks and mortar are only the beginning, that real and lasting change will involve not just physical improvements but more important, and much tougher to achieve, altered attitudes.

"This was one of the poorest neighborhoods in New York City," Goodman noted that afternoon near the stadium. "There's a thirty-year journey ahead. But the change in attitude is what's important. How we look at a place defines how desirable it is. All those sun courts, those cross-ventilated apartments, those eat-in kitchens—they were always there, but now they're encouraging people to say, 'I'm staying.'"

"Think about why a place is resonant," he added.

> It's not just the Art Deco buildings. In the first half of the twentieth century, the people who lived in this neighborhood did so by choice. They wanted to move here, and they wanted to stay. They were proud to live in the West Bronx. They weren't always prosperous, especially during the Depression, but they were optimistic. By contrast, in the second half of the twentieth century, the people who lived here didn't do so by choice, and they didn't want to stay. They felt trapped. And that was one of the many reasons this was such an unhappy community.

"Just because a place is no longer what it was doesn't mean that it's obsolete," Goodman concluded. "People who live here today are no less deserving than the people who lived here nearly a century ago. They want to feel welcome in their neighborhood. They want to feel proud. They want to be able to say to themselves, 'I am worth something because look at what I can call home.'"

Sources

When it comes to tracing the history of New York City beyond Manhattan, the Bronx is something of a stepchild compared to the lofty borough of Brooklyn, which has benefited from its larger size and its onetime status as a separate metropolis. Nonetheless, in both fiction and nonfiction, there is no lack of material with which to reconstruct the Bronx's past and the lives lived within its borders.

The largest and best repository of material is the Bronx County Historical Society (BCHS), whose offices are on Bainbridge Avenue, just north of the Grand Concourse. The society's vertical files, which are arranged by individual, location, and subject matter, contain many primary-source documents—some arcane, all fascinating—along with voluminous newspaper clippings. The *Bronx County Historical Society Journal*, the organization's quarterly publication, is valuable for investigating specific topics. Another useful research tool, especially for understanding the texture of Bronx life, is the series of profusely illustrated books covering the borough's history from its earliest days to the mid-twentieth century, published by the society and written or co-written by Lloyd Ultan, the borough's longtime historian. The society also owns the single most extensive collection of photographs of the Bronx.

The Museum of the City of New York (MCNY) is home to a priceless collection of documents, photographs, and images pertaining to Louis Risse and the planning and development of the Grand Concourse. Oral histories conducted by the Bronx Institute Oral History Project, housed at the Leonard Lief Library of Lehman College of the City University of New York, were helpful in re-creating the street's early years, as were interviews conducted by the Bronx African-American History Project at Fordham University. The Fordham project, which was started in 2002 by Mark Naison, a professor of history and African American studies at the school and director of its urban history program, has performed the invaluable service of giving voice to Bronx residents whose contributions and even presence have often gone unreported and unrecognized. The interviews compiled through this project offer the most intimate and detailed record of what it was like to be a minority resident of the Bronx.

Because the Grand Concourse and its environs are so rich in significant buildings, the designation reports issued by the New York City Landmarks

Preservation Commission are major sources of historical detail and aesthetic description, as are similar reports by the New York State Historic Preservation Office, which designated the boulevard as a State Historic Preservation District. The *AIA Guide to New York City*, written by Norval White and the late Elliot Willensky, has been described as the most comprehensive guide to the Art Deco buildings of the West Bronx, and the compliment is deserved.

The morgue of the *New York Times*, happily available electronically, is a resource of incalculable value. Also useful in charting the borough's shifting fortunes were articles from the *New York Daily News* and especially the *New York Post* and the *Bronx Home News*, the two newspapers that covered the borough most thoroughly and had the most devoted readership. Though for the most part not cited individually in the bibliography, newspaper articles on the changing fortunes of the Bronx were invaluable in reconstructing the history of the Grand Concourse and the borough as a whole.

Some of the most evocative descriptions of life on and off the Grand Concourse are found in fiction and memoir by sons and daughters of the Bronx, notably Joseph Berger, Jerome Charyn, Mary Childers, Avery Corman, Laura Shaine Cunningham, E.L. Doctorow, Nora Eisenberg, Vivian Gornick, Sherwin B. Nuland, Cynthia Ozick, and Maureen Waters. Chief among invaluable nonfiction works is *The Bronx*, by Evelyn Gonzalez, a professor at William Paterson University in New Jersey; her meticulously researched history is particularly strong in charting early development patterns. Jill Jonnes's *South Bronx Rising: The Rise, Fall, and Resurrection of an American City* offers a thorough examination of the borough's troubled past and recent efforts to restore it to health. Roberta Brandes Gratz's *The Living City: How America's Cities Are Being Revitalized by Thinking Small in a Big Way* explores many of the same themes, as does Alexander von Hoffman's *House by House, Block by Block: The Rebirth of America's Urban Neighborhoods*, which includes an extensive discussion of redevelopment efforts in the South Bronx.

The writings of the urban planner Sam Goodman—both his master's thesis and two articles in the *BCHS Journal*—provide a dry-eyed view of the rise and fall of the area affectionately known as the Golden Ghetto. The voluminous studies by Robert Stern and associates, which capture in dazzling detail the face of the city at four critical points in its history, provide thoughtful discussions of the development of the Bronx in the context of citywide growth. The *Encyclopedia of New York City*, edited by Kenneth T. Jackson, is a research tool that is, or should be, on the desk of everyone who writes about urban affairs.

Robert A. Caro's biography of master builder Robert Moses is required reading not just for an understanding of the scope and impact of the Cross Bronx Expressway but for a sense of how the twentieth-century metropolis evolved; recent rethinking of Moses's role by Columbia University professors Hilary Ballon and Kenneth Jackson, among others, in no way lessens the importance of Caro's groundbreaking achievement. It is impossible to understand the assimilated yet unexpectedly passionate Jews of the West Bronx without the insights of the historian Deborah Dash Moore, whose 1981 book *At Home in America: Second Generation New York Jews* is the single most detailed and readable study of this population. *A History of Housing in New York City* by Richard Plunz, a professor at Columbia's School of Architecture and director of the school's urban design program, offers the most comprehensive analysis of the evolution of the Bronx apartment house and other aspects of the borough's development.

Two publications by the Bronx Museum of the Arts—*Building a Borough: Architecture and Planning in the Bronx* and *Devastation/Resurrection: The South Bronx*—bring together a cornucopia of trenchant discussions from various perspectives about the borough past and present.

The writings of Marshall Berman, a professor of political science at the City University of New York, and Leonard Kriegel, formerly associated with CUNY, provide an almost novelistic take on the changes that roiled the borough in the mid- and late twentieth century. Though Mindy Thompson Fullilove's groundbreaking work *Root Shock: How Tearing Up City Neighborhoods Hurts America and What We Can Do about It* focuses on the trauma wrought by urban renewal, her analysis offers insight into the psychological impact of the social and demographic changes that washed over the Bronx through other sources of disruption. The operation collectively known as Back in the Bronx, a sprawling enterprise led by Steven M. Samtur that comprises magazines, books, and an immensely popular Web site, offers vivid firsthand accounts of what it was like to be a child of midcentury Bronx—plus, where else can you buy a chess set that includes replicas of Yankee Stadium and Loew's Paradise?

INTRODUCTION

For snapshots of life on and off the Grand Concourse in the mid-twentieth century and beyond, I drew heavily on three of my own articles. "Better Days" and especially "A Street of Dreams," which in turn drew heavily on recollections of people associated with Philly's candy store, were published in the *New York Daily News*, and "The Memory Maker," about the photography of Kourken Hovsepian—Mr. Kirk to the thousands of Bronx residents he photographed—was published in the *New York Times*. The portrait of the Bronx Slave Market was drawn largely from the writings of the crusading journalist Marvel Cooke, primarily her first article on the subject, written with Ella Baker and published in the *Crisis* in 1935. Cooke's descriptions were augmented by the recollections of Robert Gumbs, a Bronx resident whose mother took part in the market.

CHAPTER 1:
"A DRIVE OF EXTRAORDINARY DELIGHTFULNESS"

Louis Risse, the creator of the Grand Concourse, is a fascinating historical figure, one whose role has never been explored in detail. The main sources of information about his life and work are his own writings, notably the pamphlet entitled *The True History of the Conception and Planning of the Grand Boulevard and Concourse in the Bronx*, published in 1902 (issued a few years earlier under the title *History in Brief of the Conception and Establishment of the Grand Boulevard and Concourse in the 23d and 24th Wards*); "Always in My Heart: The Builder of the Grand Concourse," an article by his granddaughter, Marion Risse Morris, published in the *BCHS Journal* in 1980; and articles in the *New York Herald Tribune* and the *Bronx Home News* on the occasion of his death and on subsequent anniversaries of his passing. Although I could not locate any of Risse's American descendents, a historian from his native village in Alsace-Lorraine has been preparing a biography of the town's most famous son.

Risse's writings give a lively sense of the man. But the most striking portrait of Risse and his colleagues at work during the late nineteenth century, both in the office and in the field, emerges from a series of photographs in the collection of the Museum of the City of New York, along with Risse's drawings of the Grand Concourse, also at the museum. The two large-scale maps delineating the exact route of the boulevard are held by the office of the Bronx borough president, and the original of the Risse map that was exhibited in Paris in 1900 can be found at the New York Public Library.

The exuberant mood of the Bronx at the tail end of the nineteenth century, when the metropolis of Greater New York was poised to take shape, is captured with particular gusto by the essays in *The Great North Side, or Borough of the Bronx*, published in 1897 by the North Side Board of Trade, the primary Bronx civic group of the day. The vertical files of the BCHS are rich in material about Louis Haffen, the borough's first president and one of the more intriguing and in some respects tragic local figures of the day. The achievements of landscape architect Frederick Law Olmsted are described most completely in Witold Rybczynski's beautiful biography, *A Clearing in the Distance*.

So prolonged, controversial, and tumultuous was the planning and construction of the Grand Concourse that the *New York Times* and other publications covered the project's fortunes in detail. The bumpy but colorful history of the Heinrich Heine Fountain is recounted especially well in Michele H. Bogart's *The Politics of Urban Beauty*, in itself a comprehensive discussion of how art and politics have entwined and clashed in the city over the years.

CHAPTER 2: "GET A NEW RESIDENT FOR THE BRONX"

Theodore Dreiser's fiction and nonfiction suggests the nature of life just north of Manhattan during the early years of the twentieth century, as do oral histories conducted by the Bronx Institute Oral History Project at Lehman College. A remarkable series of photographs preserved by the BCHS—hundreds of images taken by the city's bureau of highways during the seven years the Grand Concourse was under construction—also convey a sense of the West Bronx during this era. Despite a numbing sameness, these images offer concrete visual evidence that the Bronx's great boulevard did not acquire its grandeur overnight.

Irving Howe's masterful account of the life and times of the city's Eastern European Jews, *World of Our Fathers*, provides the most extensive and expansive portrait of this population and their laborious journey from the Lower East Side to the Bronx and other parts of the city. Robert Stern's encyclopedic *New York 1930* charts the evolution of Bronx housing during the early decades of the twentieth century; more close-in accounts of the urbanization of the borough can be found in *Perception of Housing and Community: Bronx Architecture of the 1920s*, written by Brian Danforth and Victor Caliandro and published by the Hunter College Graduate Program in Urban Planning, and *The Bronx Apartment House*, written by Michael Cheilik and David Gillison and published by Lehman College.

Logan Billingsley was among the more outsized figures associated with the growth of the West Bronx, and both his wild early years and his subsequent career as a developer are brought to life by the hundreds of documents, photographs, and newspaper articles collected by his son, Robert, who also shared personal recollections of his father. *Stork Club: American's Most Famous Nightclub and the Lost World of Café Society*, Ralph Blumenthal's biography of Logan's celebrated younger brother Sherman, helped flesh out a portrait of the Billingsley family. As a major event in the Bronx, the creation of the Theodore Roosevelt was well documented by newspapers of the day and by extensive promotional material.

Of the many seminal figures who left a mark on the Grand Concourse, few were more inspiring than Andrew Thomas, in many respects the father of the garden apartment. Plunz's history of housing in New York City explores the importance of Thomas's role, as does Brian Danforth's article "Cooperative Housing," published in the *BCHS Journal*. The original prospectus for Thomas Garden and other documents related to the project are at Avery Architectural and Fine Arts Library at Columbia.

Given the primacy of the Lewis Morris apartment complex, practically everyone who ever lived there and many who did not have stories to tell about its allure; former residents whose recollections were especially useful include Harrison J. Goldin, the late David Halberstam, Max Wilson, and Joyce Sanders, author of an affectionate memoir about the building that was published by Back in the Bronx.

CHAPTER 3: "I WAS LIVING IN 'A MODERN BUILDING'"

Although the creation of the style subsequently known as Art Deco is a murky and often disputed tale, Arthur Chandler's article "The Art Deco Exposition," published in *World's Fair* magazine, offers a comprehensive description of the 1925 world's fair in Paris, the event most closely associated with the emergence of this approach to design and the decorative arts. And although volumes have been written about the style's impact in New York City, one of the most eloquent discussions of the subject is found in David Garrard Lowe's *Art Deco New York*. Don Vlack's *Art Deco Architecture in New York: 1920–1940* is among the few works on the style's presence in the city that does not give short shrift to neighborhoods outside Manhattan.

Brian Danforth's article "High-Style Architecture: Art Deco in the Bronx," published in the *BCHS Journal*, is a useful study, as are entries about specific buildings in the *AIA Guide*. But the document that most completely describes this world and brings together a considerable amount of fresh research and previously uncollected material is *Bronx Art Deco Architecture: An Exposition*, written by Donald Sullivan and Brian Danforth and published in 1976 by the Hunter College Graduate Program in Urban Planning as the catalogue to an exhibition. Along with thumbnail sketches of the major Art Deco architects of the day, the publication includes a list of hundreds of Art Deco buildings in the West Bronx and offers a detailed account of their creation, an eloquent defense of their importance, and a passionate call for action to save them from the decay lapping at their edges.

Marshall Berman also writes compellingly about these buildings in *All That Is Solid Melts into Air: The Experience of Modernity*. As a child of the West Bronx, Berman was particularly well positioned to experience firsthand the power of these apartment houses during their heyday, to mourn their deterioration, and to understand the complexity of the forces that led to the area's decline.

In piecing together the career and accomplishments of Horace Ginsbern, the premier Art Deco architect of his day, his firm's papers, held by Columbia's Avery library, is an unparalleled resource, especially useful because the documents are so well catalogued and indexed and the library's staff is so helpful. Both the firm and its founder are long gone, but the recollection of Ginsbern's grandson, John Ginsbern, whose father worked in the family firm, were invaluable in creating a warts-and-all portrait of this immensely gifted man. The personality and accomplishments of Ginsbern's longtime partner, the late Marvin Fine, and the working relationship between the two, were fleshed out by friends with firsthand memories of the pair.

Many historians and social scientists have commented on and analyzed the peculiar affinity of Jews for apartment living; in addition to Plunz, they include Andrew R. Heinze in his book *Adapting to Abundance: Jewish Immigrants, Mass*

Consumption, and the Search for American Identity; Marshall Sklare in his article "Jews, Ethnics, and the American City," published in *Commentary*; and Louis Wirth in his article "Urbanism as a Way of Life," published in the *American Journal of Sociology*.

CHAPTER 4:
"SOMETHING THAT EVERYBODY HAD IN AWE"

Re-creating the rhythms of life on and off the Grand Concourse in the middle years of the twentieth century proved an unexpectedly challenging task. Minds weaken, recollections grow vague, and many memories have an almost generic sameness. Perhaps not surprisingly, some of the freshest and most intense evocations of this time and place come from novelists and memoirists who lived on or near the boulevard, notably Jerome Charyn, Avery Corman, Laura Shaine Cunningham, and E.L. Doctorow.

The differences between the Grand Concourse and Eastern and Ocean parkways, all three of which served as crucial destinations for the city's second-generation Jews, are detailed by Allan B. Jacobs, Elizabeth Macdonald, and Yoda Rofé in *The Boulevard Book: History, Evolution, Design of Multiway Boulevards*, by Rybczynski in his biography of Olmsted, and by Plunz in his study of New York City housing. The recollections of two former borough presidents, Robert Abrams and Herman Badillo, give detail to the boulevard's longtime political role. Paul Goldberger's article in the *New York Times Magazine*, "Utopia in the Outer Boroughs," underscores the intensely public nature of this world.

Diane L. Linden's article "Ben Shahn's Murals for the Bronx Central Post Office," published in *Magazine Antiques*, describes the controversy that surrounded Shahn's original plans for the building, and Sylvie Moore's article "Bronx County Building," published in the *BCHS Journal*, gives a full account of that building's conception and construction. Laura Cunningham, who as a child lived just off the Grand Concourse, evokes the haunting presence of Joyce Kilmer and Franz Sigel parks in her memoir *Sleeping Arrangements*.

Of the countless books that have been written about the New York Yankees, the most helpful in reconstructing the team's history and examining the club's impact on the Bronx was Neil Sullivan's *The Diamond in the Bronx: Yankee Stadium and the Politics of New York*. Though largely an examination of political and economic issues associated with the stadium, Sullivan's work goes well beyond that mission to evoke splendidly the environment in which the stadium was born, to articulate the immense significance of the team's presence to the neighborhood, and to analyze the forces that led to the area's decline.

Few events in the history of sport have been covered as extensively or as colorfully as the team's first game in the stadium, on April 18, 1923. Geoffrey C. Ward's *Baseball: An Illustrated History*, written with Ken Burns, brings glowingly to life the emotions that surrounded the nation's premier team. And in *Sleeping Arrangements*, Laura Cunningham paints a picture of what it was like to live almost literally in the stadium's shadow.

As the first and only luxury hotel in the Bronx, the construction of the Concourse Plaza Hotel the same year as the arrival of the stadium was a major story for such publications as the *Bronx Home News* and the *Bronx in Tabloid*. The hotel's own promotional materials, notably a brochure entitled "Thirty Minutes from Wall Street," captured the excitement that greeted the hotel's opening. Jill Jonnes, whose affection for the hotel is evident in her book *South Bronx Rising*, details the Concourse Plaza's shifting fortunes, and the documents she amassed in connection with her research have been collected in a scrapbook that is at the Lehman College library.

The thousands of photographs taken by the Armenian immigrant Kourken Hovsepian that record six decades of domestic and celebratory life in the West Bronx are in the possession of his daughter, Mimi Vang Olsen. It was she who recounted the arc of her father's career and directed me to the former Betty Kanganis and her husband, Nick Raptis, whose wedding was one of thousands photographed by Hovsepian and celebrated at the Concourse Plaza.

The history of the Andrew Freedman Home and its remarkable founder was pieced together from material in a voluminous scrapbook on the home provided by James Crocker, executive director of the Mid-Bronx Senior Citizens Council, the organization that currently owns the building. Freedman's career and the fortunes of the institution he founded were also well covered by the newspapers of the day, though with less scrutiny than might have been merited. Geoffrey T. Hellman's article "The Bronx Palace," published in 1933 in the *New Yorker*, and Vivian Gornick's article, "A Splendid and Bitter Isolation," published nearly half a century later in the *Village Voice*, when the home still retained elements of its old grandeur, also provided a sense of the institution's singular role.

The rise and fall of Jack Molinas, another fascinating and ultimately tragic figure whose life is threaded through the Grand Concourse, is recounted most completely in Charley Rosen's biography, *The Wizard of Odds: How Jack Molinas Almost Destroyed the Game of Basketball*. Edgar Allan Poe's time in the Bronx is the subject of many biographies, and the history of the cottage where he lived and the park in which the house is located are described in most detail in Irmgard Lukmann's article "A History of Poe Park," published in the *BCHS Journal*. Marty Friedman's article "The Parkway All-Stars," published in 1981 in *New York*, casts a bright light on the generation of gifted girls and boys who came of age during the 1940s and 1950s on Mosholu Parkway, at the northern tip of the Grand Concourse, as did recollections by Bernard Gwertzman, a former *New York Times* foreign editor and a onetime member of that crew.

CHAPTER 5:
"AN ACRE OF SEATS IN A GARDEN OF DREAMS"

Although volumes have been written on the golden age of the picture palace, arguably the two finest accounts are Ben M. Hall's *The Best Remaining Seats* and David Naylor's *Great American Movie Theaters*. *Cinema Treasures: A New Look at Classic Movies Theaters*, by Ross Melnick and Andreas Fuchs, is valuable in its own right and has also spawned a Web site that bubbles with memories of Loew's Paradise and virtually every other American movie theater of note.

The late theater historian Michael Miller wrote extensively about Loew's Paradise, and his lavishly illustrated articles "Theatres of the Bronx," published in *Marquee* magazine, and "Loew's Paradise in the Bronx," published in the *Theatre Historical Society of America Annual*, are essential to an understanding of this particular landmark. Richard Stapleford's *Temples of Illusion: The Atmospheric Theaters of John Eberson*, published in connection with an exhibition at Hunter College, is a rich source of information, as is Jane Preddy's article "Glamour, Glitz and Sparkle: The Deco Theatres of John Eberson," also published in the *Theatre Historical Society of America Annual*. Surviving Eberson family members, including Suzanne Callahan and Maury Brassert, provided biographical information about their famous relative. The Eberson archives are located at the University of Pennsylvania.

Virtually everyone who lived in the West Bronx has glistening memories of Loew's Paradise, along with almost preternaturally clear recollections of other beloved local theaters, such as the Ascot, in its day one of the city's few art houses. The recollections of onetime Loew's usher Gerald McQueen, now president of the Jacob K. Javits Convention Center, provided another perspective on the borough's most famous movie house.

CHAPTER 6:
"BY THE WATERS OF THE GRAND CONCOURSE"

To understand the texture of Jewish life in the Bronx in the mid-twentieth century, the archive of *Commentary* magazine, the premier publication on American Jewish affairs during that period, is an invaluable resource, especially the articles by Ruth Glazer (later Ruth Gay), notably "West Bronx: Food, Shelter, Clothing: The Abundant Life Just Off the Grand Concourse" and Isa Kapp's "By the Waters of the Grand Concourse: Where Judaism Is Free of Compulsion." These two pieces, both published in 1949, are among the most quoted works about this milieu. Ruth Gay's memoir, *Unfinished People: Eastern European Jews Encounter America*, illuminates many aspects of the world beyond the Grand Concourse, as does Irving Howe's essay "A Memoir of the Thirties," included in *Steady Work: Essays in Politics and Democratic Radicalism*.

Deborah Dash Moore's *At Home in America* explores in depth and with much specific reference to the West Bronx the oddly secular Jewishness of the Grand Concourse, a theme she discussed further in interviews. The eras that preceded and followed the one on which Moore focused are examined by Beth S. Wenger in her book *New York Jews and the Great Depression: Uncertain Promise* and by Eli Lederhendler in his book *New York Jews and the Decline of Urban Ethnicity*.

Morton Reichek, a former journalist who grew up on the Grand Concourse, has written on his blog about the synagogues of his youth; a rich portrait of these institutions can also be found on the Web site Remembrance of Synagogues Past: The Lost Civilization of the Jewish South Bronx, laden with memories about virtually every Jewish house of worship in that part of the city. The YIVO Institute for Jewish Research is home to many books and documents that bring to life the world of the West Bronx Jews, among them the souvenir journal of the Sisterhood of Tremont Temple, published on March 23, 1929.

The Irish of the West Bronx are described by Maureen Waters in her exquisite memoir, *Crossing Highbridge*, and the local Italians by Gil Fagiani, a writer and social worker, particularly in such writings as "Stirrings in the Bronx," published in *Forward Motion*. Two articles that capture the frenzy that swirled about the Italian American boy who claimed to have seen the Virgin Mary on a rocky site just off the Grand Concourse are John T. McGreevy's "Bronx Miracle," published in *American Quarterly*, and Peter Duffy's "The Boy Who Saw the Virgin," published in the *New York Times*.

Details about the African American community on the Grand Concourse, of which little has been written, were pieced together from interviews with Mary Engelhardt, whose uncle and aunt were early residents of the street, along with interviews conducted by the Bronx African-American History Project. Information about Marvel Cooke and the Bronx Slave Market was culled from newspaper articles about the market and from Cooke's own writings, including articles in the *Crisis* and the *Compass* that were published over a fifteen-year period. Also extremely useful in reconstructing Cooke's career, her passion for her work, and the obstacles she faced was an extensive interview conducted in 1989, eleven years before her death, by Kathleen Currie for the Washington Press Club Foundation as part of the organization's oral history project for women in journalism.

CHAPTER 7:
THE GRAND CONCOURSE OF THE IMAGINATION

When I first began exploring the history of the Grand Concourse, long before the invention of the Internet, I accidentally stumbled across Eliot Wagner's 1954 novel *Grand Concourse* in the stacks of Columbia University's Butler Library. Today, the book is located within seconds via Google. But then, as now, the novel struck me as a quintessential and revealing artifact of its time, and to find the author alive and well in upstate New York and eager to discuss his literary career was a thrilling experience. That he had saved the reviews of both *Grand Concourse* and his subsequent novels was added good fortune.

E.L. Doctorow, whose short story "The Writer in the Family," is included in his collection *Lives of the Poets*, also discussed in an interview the genesis of his fiction set in the West Bronx, as did Avery Corman, whose memories were augmented by a several-hour-long stroll down the boulevard and by his article "What's Happening on the Street of Dreams," published in 1988 in the *New York Times Magazine*. Jerome Charyn's gently fictionalize memoirs make for glorious reading, as does virtually all his writing, including "Portrait of the Artist as a Young Anteater (in the Bronx)," an article on his childhood and his development as a writer that appeared in the *Washington Post.*

Arthur Kober died in 1975 and Milton Kessler in 2000, but interviews with Kessler's widow, Sonia, and his daughter, Paula, helped give a sense of the man, as did various articles about his work, including "Walking the Grand Concourse with Milton Kessler: A Critical Reception and Interview," by Adam Schonbrun and A.D. Jones, published in *Studies in American Jewish Literature*, and "Poet Kessler's Generous Spirit, Talent Remembered: Memories of 'a Great Teacher,'" by Camille Paglia, published by Binghamton University, the institution where he taught until his death. Jacob M. Appel's short story "Grand Concourse," which was brought to my attention by Andre Aciman, one of Appel's professors, was published in the *Threepenny Review* in the spring of 2007.

CHAPTER 8: "THE BOROUGH OF ABANDONMENT"

The story of the troubles that beset the Bronx during the middle and late twentieth century has been told and retold so often that it is easy to become inured to how terrible conditions were during those years, how many people were affected, and how much they suffered. And because the Grand Concourse itself was slightly less battered than some of the surrounding areas, it is tempting to assume that the boulevard escaped relatively unscathed. As writings of the period make depressingly evident, that was not the case.

Starting in the mid-1960s, local newspapers began taking note of the worsening fortunes of the borough's southern half. Alan Adelson's account of the travails of a single block along the Grand Concourse, "Living with Fear: How Crime Terrorizes a Once-Peaceful Block in New York's Bronx," which was published in 1969 in the *Wall Street Journal*, describes in intimate detail how terrifying life in the West Bronx had become, especially for the elderly.

One of the most in-depth discussions of the building abandonment that had begun eating away at the Grand Concourse is contained in Gelvin Stevenson's article "The Abandonment of Roosevelt Gardens," published in *Devastation/Resurrection*. As an economist, Stevenson was particularly well suited to the task of analyzing the complex and often murky fiscal underpinnings of the phenomenon of abandonment. Also immensely useful were the original and much longer version of Stevenson's article, his notes from interviews with the owners of the complex, and—miraculously saved after several decades—the original of the accident report for the death of Missy Holden, the eight-year-old who was beheaded in the building's elevator as she was going up to her apartment. Herbert E. Meyer's article "How Government Helped Ruin the South Bronx," published in *Fortune* in 1975, takes a broader look at the public policies that led to the collapse of the South Bronx, an area that by then included the entire lower half of the Grand Concourse.

Recent books on the urban crisis have deepened and broadened our understanding of the roots of the problems and have helped make clear how complex and perhaps intractable these problems are. Notable among thoughtful analytical studies are Thomas J. Sugrue's *The Origins of the Urban Crisis: Race and Inequality in Postwar Detroit*, which despite its focus on a city in the Midwest does much to illuminate forces at work in the Bronx, and Ira Katznelson's *When Affirmative Action Was White: An Untold History of Racial Inequality in Twentieth-Century America*, which traces the roots of many of the problems faced by American blacks back to governmental policies put in place decades earlier.

Also valuable in understanding the battering that many New York City neighborhood endured during those years is a special issue of *Dissent* magazine, published in 1987, which includes articles by a variety of sophisticated analysts, among them longtime *New York Times* architecture critic Ada Louise Huxtable. A seemingly endless series of reports issued by public and private agencies paints a grim picture of the changes washing over the Bronx during those years, as does New York City's Master Plan, which was published in 1969 and offered what in retrospect proved to be a prescient and heartbreaking analysis of the borough's troubles.

Others whose firsthand accounts give a vivid sense of life during those years are Christopher Rhoades Dykema, a social worker assigned to the area; Mary Childers, who describes her Bronx childhood in her memoir *Welfare Brat*; Adrian Nicole LeBlanc, author of the book *Random Family: Love, Drugs, Trouble, and Coming of Age in the Bronx*; and Allen Christopher Jones Jr. in his memoir *The Rat That Got Away*, the first full-length work to grow out of the Bronx African-American History Project. Though Tom Wolfe was not a child of the Bronx, he describes its horrors memorably in what has become a classic of writing on urban nightmares, *The Bonfire of the Vanities*.

CHAPTER 9: WHO KILLED THE CONCOURSE?

The fortunes of Freedomland, the charming but doomed amusement park built on the marshlands of the northeast Bronx, were extensively reported by the newspapers of the day, as was the arrival of the vast apartment complex that replaced it and the complex's subsequent impact on the West Bronx. Early coverage of Co-op

City also recalls the optimism that accompanied the opening of the project. Samuel Bleecker's book *Politics of Architecture: A Perspective on Nelson A. Rockefeller* traces the history of the complex, and the article "Co-op City: Learning to Like It," written by Denise Scott Brown and Robert Venturi and published in *Progressive Architecture* in 1970, is one of the few pieces of criticism offering kind words for the development that came to be so reviled. Helen Schwartz, one of the first residents, provided recollections of Co-op City's early days, as did Inbal Haimovich, project director of JASA Senior Services of Co-op City, and Rozaan Boone, editor of the co-op's longtime newspaper. The newspaper's early issues also reflect the initial sense of excitement that surrounded the project.

The role of Robert Moses's Cross Bronx Expressway has been analyzed in even greater detail. In an interview a few years ago, Robert Caro defended the thesis that he presents in his biography of Moses, that the highway sounded the borough's death knell. In the collection of essays *Robert Moses and the Modern City: The Transformation of New York*, Hilary Ballon and Kenneth Jackson offer another interpretation of Moses's role in the Bronx, one echoed by Ray Bromley both in the book and in his article "Not So Simple! Caro, Moses, and the Impact of the Cross-Bronx Expressway," published in 1998 in the *BCHS Journal*. Perhaps the best reconciliation of the two divergent points of view comes from Marshall Berman. In *All That Is Solid Melts into Air*, he reluctantly concludes that though the expressway may not have been the single most villainous force in the Bronx, for complex reasons that are still being understood, nothing was the same after its arrival.

In analyzing the role of the press during those years, especially valuable were the recollections of the Bronx-born city planner Sam Goodman, whose family moved from the Grand Concourse just a few months after the article bearing the headline "Grand Concourse: Hub of Bronx Is Undergoing Ethnic Changes" appeared in the *New York Times* in July 1966. Jill Jonnes, in *South Bronx Rising*, also discusses the role of the press in relationship to the exodus from the West Bronx.

CHAPTER 10: "BENDS IN THE ROAD"

For snapshots of life on and off the Grand Concourse in the fall of 2005, I relied heavily on a series of public programs and related events sponsored by the Bronx Museum of the Arts, held from September 21 to 24 as preparation for celebrating the boulevard's centennial. Interviews with Michael Bongiovi and other Bronx residents who attended the events suggested the attractions of the twenty-first-century Grand Concourse. Conversations with Ned Kaufman, the founder of Place Matters, whose specialty is historic preservation, helped clarify the changing ways people have regarded the Grand Concourse and the West Bronx over the years. Previous interviews with Millie Lopez, the late Donald Sullivan, and others, conducted when I was researching my articles about the Bronx for the *New York*

Daily News, were a reminder that even in the late 1970s and early 1980s, some people had faith in the area's future.

In describing recent improvements on and near the boulevard, Jill Jonnes's book was helpful, as was Alexander von Hoffman's *House by House, Block by Block: The Rebirth of America's Urban Neighborhoods,* along with newspaper articles about individual projects. Some of the coverage is awfully optimistic, given the many problems the Bronx continued to face as of the early twenty-first century, but as Wilhelm Ronda of the Bronx borough president's office acknowledged in an interview, not since Al Smith and the arrival of the first Yankee Stadium has construction in the West Bronx aroused such excitement.

Interviews with Bronx residents who have taken part in the Bronx African-American History Project at Fordham, some of which were shared at the events at the Bronx Museum of the Arts, are a reminder that, as Brian Purnell of the project points out, "The Grand Concourse was not grand for everyone." And Sam Goodman's analysis of the future of the West Bronx, spelled out in a series of conversations about the borough of his birth, offered a road map for the long journey ahead.

Bibliography

BOOKS

Auletta, Ken. *The Streets Were Paved with Gold*. New York: Random House, 1979.

Ballon, Hilary, and Kenneth T. Jackson, editors. *Robert Moses and the Modern City: The Transformation of New York*. New York: Norton, 2007.

Baylor, Ronald H. *Neighbors in Conflict: The Irish, Germans, Jews, and Italians of New York City: 1929–1941*. Baltimore: Johns Hopkins University Press, 1978.

Baylor, Ronald H., and Timothy Meager, editors. *The New York Irish*. Baltimore: Johns Hopkins University Press, 1996.

Behn, Noel. *Lindbergh: The Crime*. New York: Atlantic Monthly Press, 1994.

Berger, Joseph. *Displaced Persons: Growing Up American after the Holocaust*. New York: Scribner, 2001.

———. *The World in a City: Traveling the Globe through the Neighborhoods of the New New York*. New York: Ballantine Books, 2007.

Berman, Marshall. *All That Is Solid Melts into Air: The Experience of Modernity*. New York: Simon and Schuster, 1982.

Birmingham, Stephen. *The Rest of Us: The Rise of America's Eastern European Jews*. Boston: Little, Brown, 1984.

Bleecker, Samuel. *Politics of Architecture: A Perspective on Nelson A. Rockefeller*. New York: Rutledge, 1983.

Blumenthal, Ralph. *Stork Club: America's Most Famous Nightspot and the Lost World of Café Society*. Boston: Little, Brown, 2000.

Bogart, Michele H. *The Politics of Urban Beauty: New York and Its Art Commission*. Chicago: University of Chicago Press, 2006.

Braunstein, Susan L. *Getting Comfortable in New York: The American Jewish Home, 1880–1950*. Ed. Jenna Weissman Joselit. New York: Jewish Museum, 1991.

Caro, Robert A. *The Power Broker: Robert Moses and the Fall of New York*. New York: Knopf, 1974.

Chang, Jeff. *Can't Stop, Won't Stop: A History of the Hip Hop Generation*. New York: St. Martin's, 2005.

Charyn, Jerome. *The Black Swan*. New York: Thomas Dunne Books/St. Martin's, 2000.

———. *El Bronx*. New York: Mysterious Press, 1997.

————. *Bronx Boy: A Memoir.* New York: Thomas Dunne Books/St. Martin's, 2002.

————. *The Dark Lady from Belorusse.* New York: Thomas Dunne Books/St. Martin's, 1997.

Childers, Mary. *Welfare Brat: A Memoir.* New York: Bloomsbury, 2005.

Cook, Harry T. *The Borough of the Bronx, 1639–1913: Its Marvelous Development and Historical Surroundings.* New York: Harry T. Cook, 1913.

Corman, Avery. *The Old Neighborhood.* New York: Linden Press/Simon and Schuster, 1980.

Cunningham, Laura Shaine. *A Place in the Country: A Memoir.* New York: Riverhead Books, 2000.

————. *Sleeping Arrangements.* New York: Knopf, 1989.

Deák, Gloria. *Picturing New York: The City from Its Beginnings to the Present.* New York: Columbia University Press, 2000.

Derrick, Peter. *Tunneling to the Future: The Story of the Great Subway Expansion That Saved New York.* New York: New York University Press, 2001.

Doctorow, E.L. *Billy Bathgate.* New York: Random House, 1989.

————. *Lives of the Poets.* New York: Random House, 1984.

————. *World's Fair.* New York: Random House, 1985.

Drake, James A. *Richard Tucker: A Biography.* New York: Dutton, 1984.

Dreiser, Theodore. *The Genius.* New York: John Lane, 1915.

Dunlop, Beth. *Miami: Mediterranean Splendor and Deco Dreams.* New York: Rizzoli, 2007.

Eidus, Janice. *The War of the Rosens.* Lake Forest, Calif.: Behler, 2007.

Eisenberg, Nora. *The War at Home: A Memoir-Novel.* Wellfleet, Mass.: Leapfrog, 2002.

Federal Writers Project. *The WPA Guide to New York City: The Federal Writers' Project Guide to 1930s New York.* New York: Random House, 1939.

Fedorchak, Vincent. *Fuzz One: A Bronx Childhood.* New York: Testify Books, 2005.

Fine, Marshall. *Accidental Genius: How John Cassavetes Invented the American Independent Film.* New York: Miramax Books/Hyperion, 2005.

Freedman, Samuel G. *Who She Was: My Search for My Mother's Life.* New York: Simon and Schuster, 2005.

Fullilove, Mindy Thompson. *Root Shock: How Tearing Up City Neighborhoods Hurts America and What We Can Do about It.* New York: One World/Ballantine Books, 2004.

Gaines, Stephen. *Obsession: The Life and Times of Calvin Klein.* Secaucus, N.J.: Carol, 1994.

Gay, Ruth. *Unfinished People: Eastern European Jews Encounter America.* New York: Norton, 1996.

Gelb, Arthur. *City Room.* New York: Putnam, 2003.

Geller, Victor B. *Take It Like a Soldier.* Jerusalem: Self-published, 2007.

Gold, Michael. *Jews without Money.* New York: Horace Liveright, 1930.

Gonzalez, Evelyn. *The Bronx.* New York: Columbia University Press, 2004.

Gornick, Vivian. *Fierce Attachments.* New York: Farrar, Straus, and Giroux, 1987.

Gratz, Roberta Brandes. *The Living City: How America's Cities Are Being Revital-ized by Thinking Small in a Big Way.* New York: Simon and Schuster, 1989.

Gross, Michael. *Genuine Authentic: The Real Life of Ralph Lauren.* New York: HarperCollins, 2003.

Hall, Ben M. *The Best Remaining Seats: The Story of the Golden Age of the Movie Palace.* New York: Bramhall House, 1961.

Handlin, Oscar. *The Uprooted.* Boston: Little, Brown, 1951.

Heinze, Andrew R. *Adapting to Abundance: Jewish Immigrants, Mass Consumption, and the Search for American Identity.* New York: Columbia University Press, 1990.

Hermalyn, Gary, and Robert Kornfeld. *Landmarks of the Bronx.* New York: Bronx County Historical Society, 1989.

Howe, Irving. *Steady Work: Essays in Politics and Democratic Radicalism, 1953–1966.* New York: Harcourt, Brace and World, 1966.

———. *World of Our Fathers.* New York: Harcourt Brace Jovanovich, 1976.

Howe, Irving, and Kenneth Libo, editors. *How We Lived: A Documentary History of Immigrant Jews in America, 1880–1930.* New York: Putnam, 1983.

Huxtable, Ada Louise. *Will They Ever Finish Bruckner Boulevard? A Primer on Urbicide.* New York: Macmillan, 1970.

Israelowitz, Oscar. *Synagogues of New York City: History of a Jewish Community.* New York: Israelowitz, 2000.

Jackson, Kenneth T., editor. *Encyclopedia of New York City.* New Haven, Conn.: Yale University Press, 1995.

Jacobs, Allan B., Elizabeth Macdonald, and Yoda Rofé. *The Boulevard Book: History, Evolution, Design of Multiway Boulevards.* Cambridge, Mass.: MIT Press, 2002.

Jacobs, Jane. *The Death and Life of Great American Cities.* New York: Random House, 1961.

Jenkins, Stephen. *The Story of the Bronx, 1639–1912.* New York: Putnam's Sons, 1912.

Jones, Allen Christopher, Jr., and Mark Naison. *The Rat That Got Away: A Memoir by Allen Jones.* New York: Fordham University Press, forthcoming.

Jonnes, Jill. *South Bronx Rising: The Rise, Fall, and Resurrection of an American City.* New York: Fordham University Press, 2002.

Katznelson, Ira. *When Affirmative Action Was White: An Untold History of Racial Inequality in Twentieth-Century America.* New York: Norton, 2005.

Kessler, Milton. *The Grand Concourse: Poems by Milton Kessler.* Binghamton, N.Y.: MMS, State University of New York at Binghamton, 1990.

King, Moses. *King's Handbook of New York City*. Boston: Moses King, 1893.

Klein, Robert. *The Amorous Busboy of Decatur Avenue: A Memoir*. New York: Touchstone, 2005.

Kober, Arthur. *Thunder over the Bronx*. New York: Simon and Schuster, 1935.

———. *My Dear Bella*. New York: Random House, 1941.

Korman, Marvin. *In My Father's Bakery: A Bronx Memoir*. New York: Red Rock, 2003.

Kriegel, Leonard. *Notes for the Two-Dollar Window: Portraits from an American Neighborhood*. New York: Saturday Review/Dutton, 1976.

Kroessler, Jeffrey A. *New York Year by Year: A Chronology of the Great Metropolis*. New York: New York University Press, 2002.

LeBlanc, Adrian Nicole. *Random Family: Love, Drugs, Trouble, and Coming of Age in the Bronx*. New York: Scribner, 2003.

Lederhendler, Eli. *New York Jews and the Decline of Urban Ethnicity, 1950–1970*. Syracuse, N.Y.: Syracuse University Press, 2001.

LoBrullo, Vincent. *Stanley Kubrick, a Biography*. New York: Donald I. Fine Books, 1997.

Lowe, David Garrard. *Art Deco New York*. New York: Watson-Guptill, 2004.

Lubell, Samuel. *Future of American Politics*. New York: Harper and Brothers, 1952.

Mahler, Jonathan. *Ladies and Gentleman, the Bronx Is Burning: 1977, Baseball, Politics, and the Battle for the Soul of a City*. New York: Farrar, Straus, and Giroux, 2005.

Margolick, David. *Beyond Glory: Joe Louis vs. Max Schmeling, and a World on the Brink*. New York: Knopf, 2005.

McNamara, John. *History in Asphalt: The Origin of Bronx Street and Place Names*. New York: Bronx County Historical Society, 1991.

———. *McNamara's Old Bronx*. New York: Bronx County Historical Society. 1989.

Melnick, Ross, and Andreas Fuchs. *Cinema Treasures: A New Look at Classic Movie Theaters*. St. Paul, Minn.: MBI, 2004.

Merola, Mario, with Mary Ann Giordano. *Big-City D.A.* New York: Random House, 1988.

Moore, Deborah Dash. *At Home in America: Second Generation New York Jews*. New York: Columbia University Press, 1981.

Naylor, David. *Great American Movie Theaters*. Washington, D.C.: Preservation Press, 1987.

North Side Board of Trade. *The Great North Side, or Borough of the Bronx*. New York: North Side Board of Trade, Knickerbocker Press, 1897.

Nuland, Sherwin B. *Lost in America: A Journey with My Father*. New York: Knopf, 2003.

Ozick, Cynthia. *The Puttermesser Papers*. New York: Knopf, 1997.

Pearson, Michael. *Dreaming of Columbus: A Boyhood in the Bronx.* Syracuse, N.Y.: Syracuse University Press, 1999.

Plunz, Richard. *A History of Housing in New York City.* New York: Columbia University Press, 1990.

Rider, Fremont, editor. *Rider's New York City: A Guide Book for Travelers.* New York: Macmillan, 1924.

Rosen, Charley. *The Wizard of Odds: How Jack Molinas Almost Destroyed the Game of Basketball.* New York: Seven Stories, 2001.

Rosen, Gerald. *Growing Up Bronx.* Berkeley, Calif.: North Atlantic Books, 1984.

Rosenthal, Mel. *In the South Bronx of America.* Willimantic, Conn.: Curbstone, 2000.

Rozan, S.J. *Concourse.* New York: St. Martin's, 1995.

Rybczynski, Witold. *A Clearing in the Distance: Frederick Law Olmsted and America in the 19th Century.* New York: Scribner, 1999.

Samtur, Stephen M., and Martin A. Jackson. *The Bronx, Lost, Found, and Remembered: 1935–1975.* Scarsdale, N.Y.: Back in the Bronx, 1999.

Sharp, Dennis. *The Picture Palace and Other Buildings for the Movies.* New York: Praeger, 1969.

Simon, Kate. *Bronx Primitive: Portraits in a Childhood.* New York: Viking, 1982.

Stern, Robert A.M., David Fishman, and Jacob Tilove. *New York 2000: Architecture and Urbanism between the Bicentennial and the Millennium.* Monacelli, 2006.

Stern, Robert A.M., Gregory Gilmartin, and Thomas Mellins. *New York 1930: Architecture and Urbanism between the Two World Wars.* New York: Rizzoli, 1988.

Stern, Robert A.M., Thomas Mellins, and David Fishman. *New York 1960: Architecture and Urbanism between the Second World War and the Bicentennial.* New York: Monacelli, 1997.

Sugrue, Thomas J. *The Origins of the Urban Crisis: Race and Inequality in Postwar Detroit.* Princeton, N.J.: Princeton University Press, 1996.

Sullivan, Neil. *The Diamond in the Bronx: Yankee Stadium and the Politics of New York.* New York: Oxford University Press, 2001.

Ultan, Lloyd. *The Beautiful Bronx: 1920–1950.* New York: Arlington House, 1979.

Ultan, Lloyd, and Gary Hermalyn. *The Birth of the Bronx: 1609–1900.* New York: Bronx County Historical Society, 2000.

———. *The Bronx in the Innocent Years: 1890–1925.* New York: Harper and Row, 1985.

———. *The Bronx: It Was Only Yesterday: 1935–1965.* New York: Bronx County History Society, 1992.

Ultan, Lloyd, and Barbara Unger. *Bronx Accent: A Literary and Pictorial History of the Borough.* New Brunswick, N.J.: Rutgers University Press, 2000.

Vlack, Don. *Art Deco Architecture in New York: 1920–1940*. New York: Harper and Row, 1974.

von Hoffman, Alexander. *House by House, Block by Block: The Rebirth of America's Urban Neighborhoods*. New York: Oxford University Press, 2003.

Wagner, Eliot. *Grand Concourse*. New York: Bobbs-Merrill, 1954.

Ward, Geoffrey C., and Ken Burns. *Baseball: An Illustrated History*. New York: Knopf, 1994.

Waters, Maureen. *Crossing Highbridge: A Memoir of Irish America*. Syracuse, N.Y.: Syracuse University Press, 2001.

Wells, James L., Louis Haffen, and Josiah A. Briggs, editors. *The Bronx and Its People: A History, 1609–1927*. New York: Lewis Historical, 1927.

Wenger, Beth S. *New York Jews and the Great Depression: Uncertain Promise*. New Haven, Conn.: Yale University Press, 1996.

White, Norval, and Elliot Willensky. *The AIA Guide to New York City: Fourth Edition*. New York: Three Rivers, 2000.

Wolfe, Tom. *The Bonfire of the Vanities*. New York: Farrar, Straus, and Giroux, 1987.

ARTICLES

Adelson, Alan. "Living with Fear: How Crime Terrorizes a Once-Peaceful Block in New York's Bronx." *Wall Street Journal*, Nov. 7, 1969.

Appel, Jacob M. "Grand Concourse." *Threepenny Review* (spring 2007) (fiction).

Baker, Ella, and Marvel Cooke. "The Bronx Slave Market." *Crisis*, November 1935.

Bauer, Sherrie L. "Development of New York's Puerto Rican Community," *Bronx County Historical Society Journal* 21, no. 1 (spring 1988).

Berman, Marshall. "Roots, Ruins, Renewals: City Life after Urbicide." *Village Voice*, Sept. 4, 1984.

———. "Ruins and Reform: New York Yesterday and Today." *Dissent* (fall 1987).

———. "Views from the Burning Bridge." *Urban Mythologies: The Bronx Represented Since the 1960s* (Bronx Museum of the Arts), Apr. 8–Sept. 5, 1999.

Bromley, Ray. "Not So Simple! Caro, Moses, and the Impact of the Cross-Bronx Expressway." *Bronx County Historical Society Journal* 35, no. 1 (spring 1998).

Brown, Denise Scott, and Robert Venturi. "Co-op City: Learning to Like It." *Progressive Architecture*, Feb. 1970.

Chandler, Arthur. "The Art Deco Exposition." *World's Fair* 8, no. 3 (1988).

Charyn, Jerome. "Portrait of the Artist as a Young Anteater (in the Bronx)." *Washington Post*, June 22, 2003.

———. "The Rough Adventure of the Street." *Dissent* (fall 1987).

Corman, Avery. "What's Happening on the Street of Dreams." *New York Times Magazine*, Nov. 20, 1988.

Danforth, Brian. "Cooperative Housing." *Bronx County Historical Society Journal* 15, no. 2 (fall 1978).

———. "High-Style Architecture: Art Deco in the Bronx." *Bronx County Historical Society Journal* 14, no. 1 (spring 1977).

Fagiani, Gil. "Stirrings in the Bronx." *Forward Motion* (Oct.-Nov. 1985).

Fishman, Michael. "The Grand Concourse: Historic, Visionary, Abused and Reimagined." *New York Transportation Journal* 8, no. 1 (fall 2004).

Frazier, Ian. "Utopia in the Bronx: Co-op City and Its People." *New Yorker,* June 26, 2006.

Freedman, Morris. "The Real Molly Goldberg: Baalebosteh of the Air Waves." *Commentary* (Apr. 1956).

Friedman, Marty. "The Parkway All-Stars." *New York,* Oct. 26, 1981.

Glazer, Ruth. "The Jewish Delicatessen: The Evolution of an Institution." *Commentary* (Mar. 1946).

———. "West Bronx: Food, Shelter, Clothing: The Abundant Life Just Off the Grand Concourse." *Commentary* (June 1949).

Goldberger, Paul. "Utopia in the Outer Boroughs." *New York Times Magazine,* Nov. 4, 1984.

Goodman, Sam. "The Golden Ghetto: The Grand Concourse in the 20th Century, Part I." *Bronx County Historical Society Journal* 41, no. 1 (spring 2004).

———. "The Golden Ghetto: The Grand Concourse in the Twentieth Century, Part II, 1960 to the Present." *Bronx County Historical Society Journal* 42 (fall 2005).

———. "A Second Hundred Years on the Grand Concourse." *TGC* (Bronx Museum of the Arts) (spring-summer 2006).

Gornick, Vivian. "A Splendid and Bitter Isolation." *Village Voice,* July 16–22, 1980.

Hellman, Geoffrey T. "The Bronx Palace." *New Yorker,* Apr. 8, 1933.

Huxtable, Ada Louise. "Stumbling toward Tomorrow: The Decline and Fall of a New York Vision." *Dissent* (fall 1987).

Kapp, Isa. "By the Waters of the Grand Concourse: Where Judaism Is Free of Compulsion." *Commentary* (Sept. 1949).

Kriegel, Leonard. "In the Country of the Other: A Journey Back to the North Bronx." *Dissent* (fall 1987).

———. "Last Stop on the D Train: In the Land of the New Racists." *American Scholar* 39 (spring 1970).

———. "Synagogues: On Being a Believing Nonbeliever." *American Scholar* (autumn 2000).

Linden, Diana L. "Ben Shahn's Murals for the Bronx Central Post Office." *Magazine Antiques* (Nov. 1996).

Lukmann, Irmgard. "A History of Poe Park." *Bronx County Historical Society Journal* 18, no. 1 (spring 1981).

McGreevy, John T. "Bronx Miracle," *American Quarterly* 52, no. 3 (Sept. 2000).

Meyer, Herbert E. "How Government Helped Ruin the South Bronx." *Fortune* (Nov. 1975).

Miller, Michael. "Loew's Paradise in the Bronx." *Theatre Historical Society of America Annual* 2 (1975).

———. "Theatres of the Bronx." *Marquee* 4, no. 3 (1972).

Moore, Sylvie. "Bronx County Building." *Bronx County Historical Society Journal* 10, no. 2 (July 1973).

Morris, Marion Risse. "Always in My Heart: The Builder of the Grand Concourse." *Bronx County Historical Society Journal* 17, no. 1 (spring 1980).

Paglia, Camille. "Poet Kessler's Generous Spirit, Talent Remembered: Memories of 'a Great Teacher.'" Binghamton University, Mar.-Apr. 2000.

Paneth, Donald. "Bronx Housewife: The Life and Opinions of Mrs. Litofsky." *Commentary* (Feb. 1951).

Preddy, Jane. "Glamour, Glitz and Sparkle: The Deco Theatres of John Eberson." *Theatre Historical Society of America Annual* 16 (1989).

Rosenblum, Constance. "Better Days." *New York Daily News Sunday Magazine*, Dec. 12, 1976.

———. "The Memory Maker." *New York Times*, Apr. 14, 2003.

———. "A Street of Dreams." *New York Daily News Sunday Magazine*, Apr. 6, 1980.

Sanders, Joyce. "The Lewis Morris." *Back in the Bronx* 10, issue 41 (summer 2003).

Schonbrun, Adam, and A.D. Jones. "Walking the Grand Concourse with Milton Kessler: A Critical Reception and Interview." *Studies in American Jewish Literature* 9, no. 2 (1990).

Sklare, Marshall. "Jews, Ethnics, and the American City." *Commentary* (Apr. 1972).

Stevenson, Gelvin. "The Abandonment of Roosevelt Gardens." *Devastation/ Resurrection: The South Bronx* (Bronx Museum of the Arts) (1979).

Wirth, Louis. "Urbanism as a Way of Life." *American Journal of Sociology* 44, no. 1 (July 1938).

REPORTS, EXHIBITION CATALOGUES,
UNPUBLISHED MANUSCRIPTS

Agovino, Michael A. "Bronx Utopia: A Memoir from the Periphery of New York." Unpublished manuscript, 2006.

American Jewish Congress. *The Grand Concourse: Promise and Challenge*. American Jewish Congress, 1967.

Bronx Museum of the Arts. *Urban Mythologies: The Bronx Represented since the 1960s*, 1999.

———. *Building a Borough: Architecture and Planning in the Bronx, 1890–1940*, 1986.

———. *Devastation/Resurrection: The South Bronx*, 1979.

Cheilik, Michael, and David Gillison. *The Bronx Apartment House*. Herbert H. Lehman College, City University of New York, 1977.

———. *Public Buildings in the Bronx*. Herbert H. Lehman College, City University of New York, 1978.

Concourse Jewish Community Council. *The Invisible Jewish Poor Made Visible: An Interim Report*. Mar. 1, 1974.

Currie, Kathleen. Interview with Marvel Cooke, Washington Press Club Foundation Women in Journalism Oral History Project, Oct. 1989.

Danforth, Brian, and Victor Caliandro. *Perception of Housing and Community: Bronx Architecture of the 1920s*. Hunter College Graduate Program in Urban Planning, City University of New York, 1977.

Golden, Martha. *The Grand Concourse: Tides of Change*. New York City Landmarks Preservation Commission Scholars Program, 1976.

Goodman, Sam. "The Golden Ghetto: Bronx, New York, 1920–1980." Master's thesis, University of Bridgeport, 1981.

Haffen, Louis. *Borough of the Bronx: A Record of Unparalleled Progress and Development*, 1909.

Hart, Neil Edward. "The Sidewalks Were Red." Unpublished manuscript, collection of the Bronx County Historical Society, 1972.

Hermalyn, Gary D. "The Concourse Action Program: The Parks." Bronx County Historical Society, May 1981.

New York City Planning Commission. *Grand Concourse, Special Zoning District Proposal*, Apr. 1986.

———. *Plan for New York City: A Proposal; Vol. 2: The Bronx*, 1969.

———. *A Portrait of the Elderly*, Mar. 1974.

Risse, Louis. *The True History of the Conception and Planning of the Grand Boulevard and Concourse in the Bronx*, Dec. 8, 1902.

South Bronx Development Office. *Areas of Strength/Areas of Opportunity: South Bronx Revitalization Program and Development Guide Plans*, Dec. 1980.

Stapleford, Richard. *Temples of Illusion: The Atmospheric Theaters of John Eberson*. Hunter College, Apr. 13–May 27, 1988.

Stevenson, Gelvin. "Roosevelt Gardens: A Case Study in Housing Abandonment." Unpublished manuscript, 1978.

Sullivan, Donald, and Brian Danforth. *Bronx Art Deco Architecture: An Exposition*. Hunter College Graduate Program in Urban Planning, City University of New York, 1976.

Women's City Club of New York. *With Love and Affection: A Study of Building Abandonment*. 1977.

WEB SITES

Cinema Treasures. http://cinematreasures.org.

Forgotten New York. "Street Scenes." http://www.forgotten-ny.com/STREET SCENES/STREET SCENES HOME PAGE/strthome.html.

The Grand Concourse—The Champs Elysee of the Bronx. http://www.brorson. com/BronxWeb/GrandConcourse1.html.

Morton Reichek, Octogenarian weblog. "The Synagogues on 169th Street and Other Matters." Sept 4, 2007. http://octogenarian.blogspot. com/2007_09_01_archive.html.

Remembrance of Synagogues Past: The Lost Civilization of the Jewish South Bronx. http://www.bronxsynagogues.org.

PERMISSIONS

Esther Hoffman Beller, Interview No. 134.2 by Emita Hill. Bronx Institute Oral History Project, Special Collections, Leonard Lief Library of Lehman College, the City University of New York. Oct. 27, 1983.

Elba Cabrera, Interview No. 289 by T. Kirin, Bronx Institute Oral History Project, Special Collections, Leonard Lief Library of Lehman College, the City University of New York. Aug. 5, 1986.

Frances Rosenblatt, Interview No. 126 by D. Malone, Bronx Institute Oral History Project, Special Collections, Leonard Lief Library of Lehman College, the City University of New York. July 13, 1983.

Index

About the Author

CONSTANCE ROSENBLUM, the longtime editor of the City section of the *New York Times,* is the author of *Gold Digger: The Outrageous Life and Times of Peggy Hopkins Joyce* and the editor of *New York Stories: The Best of the City Section of the New York Times,* available from NYU Press. Previously, she was editor of the *Times's* Arts and Leisure section.